Governance for Drought Resilience

Hans Bressers · Nanny Bressers
Corinne Larrue
Editors

Governance for Drought Resilience

Land and Water Drought Management in Europe

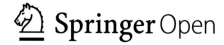

Editors
Hans Bressers
University of Twente
Enschede, Overijssel
The Netherlands

Nanny Bressers
Vechtstromen Water Authority
Almelo
The Netherlands

Corinne Larrue
Val de Marne
Université Paris Est Créteil
Marne-la-Vallée
France

ISBN 978-3-319-29669-2   ISBN 978-3-319-29671-5   (eBook)
DOI 10.1007/978-3-319-29671-5

Library of Congress Control Number: 2016931423

Printed on acid-free paper

This Springer imprint is published by SpringerNature
The registered company is Springer International Publishing AG Switzerland

Foreword

Overcoming Drought and Water Shortages with Good Governance

On June 20, 2013, a picture went viral on the Internet in which two students were seemingly sailing through the shopping streets of Enschede, my hometown and the biggest city in the region of Vechtstromen, a water authority in the east of the Netherlands. After an extraordinary downpour it was the first time ever that these streets were flooded. It was quickly forgotten that just days before this same city was regarded a risky "hotspot" of unusually high inner city temperatures after a long heat wave that caused many creeks in its rural surroundings to completely dry out. This example is consistent with a general pattern in many areas in Northwest Europe: weathers are often more extreme than they used to be and water management will have to cope with that, through increasing the resilience of our water systems. This book, based on the many insights that are gathered from the project "Benefit of governance in drought adaptation" (DROP, Interreg IVb NWE) provides an overview of a variety of drought situations in six areas in Northwest Europe, the measures taken to improve these situations and especially an in-depth treatment of the governance conditions that support or restrict the realization of these measures.

The people in Northwest Europe that have always, and rightfully so, regarded themselves as living in a water rich area not only need to get used to more weather extremes, but especially to one of the extremes: drought. The availability of sufficient water of apt quality for purposes such as agriculture, nature, and service and drinking water production has become less self-evident and will likely even become less certain in the future. Each of these three functions gets special attention in the comparative chapters in this book. Sufficient freshwater supply is also important for shipping, use, and discharge of cooling water, and for flushing waterways in the low-lying parts of the country to prevent saltwater intrusion in a country such as the Netherlands. Generally, sufficient freshwater supply is a matter of utmost importance as part of a good investment climate for economic activities that are dependent on water. Nevertheless, outside of the realm of water experts the awareness of the risk of periods of present and future water scarcity is generally low and only slowly increasing. This lack of drought awareness among citizens (voters) and landowners (farmers) makes it difficult for water managers to impose a forceful drought resilience strategy. While doing everything that is possible given this shallow support base, it is therefore essential to simultaneously pay continuous attention to awareness raising.

In this book six regions from five countries are studied: Germany, Belgium, France, the UK, and the Netherlands. While all regions share the common characteristics of the Northwest European area, such as on average sufficient water supply, a high level of economic development, and the context of European policy schemes, they also illustrate a high degree of variation in their specific situations, hydrology, water use, and institutional arrangements for water management. This variety in the situation on the ground requires tailored action.

Different packages of measures are possible to increase resilience toward drought. Three general strategies can be discerned. The first is a set of reactive measures. To these belong the transport of water from other regions or from bigger rivers to the dry areas. Something that is not only costly, but can also have negative side effects, for instance, on water quality in vulnerable nature areas. Another one is setting minimal flow requirements, for instance, to protect aquatic biotopes. While reactive it also induces more preventive measures. Among the preventive measures are all kinds of interventions to save and hold water available from wet periods, and to increase the buffering capacity of the soil and the water system. Lastly there are measures of the adaptive strategy: to accept the limitations of the natural system and consequently adapt the water use to the drought risks that the system generates. Of course, farmers could still decide to grow high value but vulnerable crops on drought-prone lands, but when yield fails every ten years or so, they should view it as an entrepreneurial risk that they cannot have avoided or compensated by public authorities. While water demand management measures like these are still unusual, it is wise to start preparing the support basis for them and develop ideas on their future implementation.

All strategies require an appropriate land use and a robust water system. The characteristics of the landscape and those of the water system should be in harmony. As with water uses the natural conditions should be put center stage. Thus, not "water levels should be adjusted to follow chosen human land and water uses," but "land and water uses should follow water levels as they result from natural circumstances." For both the resilience of the water system and the quality of the landscape and cityscape, so-called green–blue veins and grids are of crucial importance. These are not only of esthetical value, which is important in its own right, but also provide the function of improving buffering capacity for both too much and too little water. The risk is that more attention for floods leads to measures that increase vulnerability for droughts, or the other way around. But since both are occurring now with more frequency, this would be an inappropriate response. As the Somerset case and several other examples in this book show, there is certainly a need for an increased integration of flood and drought management.

A call for further integration is justified but not unproblematic. The challenges put forward to the governance of water systems are manifold and dynamic. Consequently expert knowledge on, for instance, hydrology or engineering, while still essential, is not enough. The need for a sufficient support basis by both the people involved and the financial resources needed adds to the complexity. In 1997 John Dryzek discerned in his book "The Politics of the Earth" three what he called 'discourses': leave it to the experts: administrative rationalism, leave it to the people: democratic pragmatism, and leave it to the market: economic rationalism. While a discourse is a way of framing a subject, including some and excluding other issues, using key words and stories to reinforce that way of thinking and shield it from other ways, the challenge for water governance nowadays is precisely to overcome such boundaries. Thereby the expert side is becoming broader, with also multi-actor project management skills and governance expertise. For the people's side their appreciation of the waterways in their surroundings is important, but also their sense of responsibility to guard this heritage. Water management is not only about applying knowledge in the right technocratic way, it is also about the equal consideration of interests. And for the market side water pricing for economic activities' water use, and innovative water treatment that produces energy and other resources, like nitrates, phosphates and even clean service water, contribute to the development of a blue-based economy.

Like the three 'discourses' which should be coupled rather than separated, various forms of knowledge also should be coupled. Local knowledge and practitioners knowledge is sometimes as valuable as scientific knowledge, especially when these sources are brought together. Complex and dynamic processes like water management need a continuous flow of insights into possible scenarios and into the dilemmas that need to be overcome. Many can contribute to such insights. This makes useable knowledge often 'negotiated knowledge', no ultimate or final truth, but a provisional common ground to proceed upon.

Drought resilience management requires not only such common ground, but also collaboration between several stakeholders to walk on that common ground together. This book concentrates on governance conditions as a supportive or restrictive context for drought resilience management. To a large extent it concentrates on the old but vital question: what enables productive cooperation? Its cases and comparisons provide a wealth of answers. There is not one single answer that fits every situation, and although that certainly makes matters complex, it is a complexity we should not only accept but also embrace. It is through regional tailoring, combined with learning-oriented exchange that the most optimal solutions for tackling drought can be created.

I would like to compliment the editors of this book for the insights they provide in the essence of good governance when it comes to water shortage and drought. It is of great value for water practitioners who can learn from this book how to make water management more adaptive to changing climate conditions.

Stefan Kuks
President of Vechtstromen Water Authority

Contents

Chapter 1
Introduction

Nanny Bressers, Hans Bressers and Corinne Larrue

This book is about governance for drought resilience. But that simple sentence alone might rouse several questions. Because what do we mean with drought, and how does that relate to water scarcity? And what do we mean with resilience, and why is resilience needed for tackling drought? And how does governance enter this equation? We argue that governance assessment—the study of restricting and facilitating characteristics of a governance setting—can greatly aid implementation of drought adaptation measures, thereby increasing drought resilience. In this chapter we will first discuss the occurrence of drought in (Northwest) Europe, and why governance matters for increasing drought resilience (Sect. 1.1). Second, we will shed more light on the notion of governance and how governance is dealt with by us (Sect. 1.2). Third, we will review existing governance assessment methods (Sect. 1.3). Fourth, we will use this knowledge on governance and governance assessment to introduce our key principles in governance assessment, and discuss how we have applied these in a real-life project setting (Sect. 1.4). Fifth, and last, this chapter ends with an outlook into the rest of the book, to guide the reader in reading (Sect. 1.5).

N. Bressers (✉)
Water Board Vechtstromen, Kooikersweg 1, Almelo, The Netherlands
e-mail: n.bressers@vechtstromen.nl

H. Bressers
University of Twente, Enschede, The Netherlands

C. Larrue
Corinne Larrue works at University of Rabelais de Tours, Tours, France

© The Author(s) 2016
H. Bressers et al. (eds.), *Governance for Drought Resilience*,
DOI 10.1007/978-3-319-29671-5_1

1

1.1 Introduction: Why Governance for Drought Resilience?

Drought and water scarcity are very visible and prominent problems in some areas of the world. In Europe the south of the continent is very prone to drought, and suffers great (economical) damages as a result of it. In Northwest Europe drought is generally not recognized as a major problem. Water surplus—for instance in flooding—receives much more attention and is generally the focus of water managers. However, even though drought may not be as visible as flooding, that does not mean it does not exist. For instance, the heat wave and drought of summer 2003 caused the loss of thousands of human lives in Europe and had a financial impact of 13 billion euros (COPA COGECA 2003). Agricultural production declines as a result of precipitation shortages, reduced groundwater levels and so on. Nature areas suffer from drought as well, due to drops in groundwater levels, more competition for remaining water supplies and increasing eutrophication. Freshwater management plays a vital role in both the supply–demand balance as well as the effects drought has on water quality. Thus, drought and water scarcity in Europe have an impact on agriculture, nature, freshwater—and as a result also impact the economy as a whole and people's health. In this book we will mostly discuss 'drought', but that is not to exclude water scarcity. Rather, we use 'drought' as common denominator for both the issue of drought as well as water scarcity.

Drought can occur in virtually any climatic regime, in both high and low rainfall areas. In contrast to aridity, which is a permanent feature of the climate and is restricted to low rainfall areas, drought is a temporary water shortage condition compared to an average situation. It is usually the consequence of a natural reduction in the amount of rainfall received over an extended period of time, which can be caused or aggravated by other climatic factors, such as high temperatures, high winds or low relative humidity. Drought can also be induced by human factors, causing, for instance, excessive demands over a supply–demand system. Following this, and depending on the main causes or impacts, some definitions of droughts have been proposed, which are usually grouped into four types (Wilhite and Glantz 1987):

- meteorological drought, which is mainly due to a long period of no or very low rainfall;
- hydrological drought, which is characterized by river flows that are below average;
- agricultural drought, which refers to a soil moisture deficit affecting crops;
- mega drought, which is a persistent and extended drought that lasts for a much longer period than normal.

Additionally, some authors will also consider another type called 'socioeconomic drought', which occurs when the demand for water exceeds the supply. Here, it may be important to highlight the difference between drought (and drought impacts) and water scarcity. Water scarcity and drought are two interrelated but distinct concepts. Water scarcity may result from a range of phenomena, which may be produced by natural causes, such as drought, but can also be induced by human activities only, or, as is usually the case, may result from the interaction of both (Pereira et al. 2002).

Drought is expected to increase in the future as a result of climate change. In 2007 11 % of the European population and 17 % of the European territory were affected by drought (EC 2007). Already, it can be noted that the number of people and areas in Europe affected by drought and water scarcity has increased with 20 % between 1976 and 2006 (EC 2007). The total cost of these 30 years of droughts amounts to 100 billion euros (ibid.). This makes it very important to deal with drought and water scarcity now, and to increase drought resilience before the problem grows even bigger. In the 2007 Communication the European Union clearly states that devising effective drought risk management strategies has to be regarded an EU priority.

But in order to optimize drought resilience not only the physical situation must be studied and worked on. The governance setting matters a great deal in determining the effectiveness of drought adaptation measures and facilitating their implementation. Governance ultimately revolves around the social, organizational, political and juridical dimensions and how actors operate in these dimensions to work on issues such as drought resilience. In many instances (technological) innovations for increasing drought resilience do exist, but their implementation is hampered because of factors in the governance setting. As an example, an innovation is developed and tested by a local actor, but its upscaling for broader application is limited because actors in the region do not engage in intensive networking or suffer from the 'not invented here' syndrome and the resulting lack of ownership decreases the potential to upscale. Another example is when the ideas for an innovation do exist, but resources lack to fund their proper development, as a result of little experienced urgency for change at regime level. In other words; governance matters, and governance is broader than just the (governmental) actors and their adoption of drought resilience measures.

In the upcoming chapters governance and the role of governance in drought resilience will be further discussed, translated into a fully developed governance assessment model, and applied to real-life cases. Through this discussion, development and application we want to provide assistance to practitioners working on increasing drought resilience. This book is primarily written by scientists, but strongly embedded in our interaction with practitioners. Some chapters are also co-written by these practitioners. We believe that this work on the edge of science and practice contributes to an innovative perspective on drought resilience, and is an example of the novelty and applicability of our work.

1.2 Defining Governance

Governance has been extensively discussed in political sciences and public administration literature. It is often presented as part of a more general shift from government to governance (Kooiman 1993, 2003; Klijn and Koppenjan 2000). Governance in that sense is the interaction of public and private actors aimed at solving societal problems or creating societal opportunities in an institutional context with a normative foundation (summary from Bressers 2011: 25; based on Kooiman 2003: 4). Government is thus no longer the sole decision-maker, and allows for direct influence of other parties. The advantages of this are numerous, for instance greater support from stakeholders, higher quality of work as a result of expert and layman input, and greater legitimacy of decisions (van Schie 2010: 33; Termeer 2009: 300). Another effect of the shift to governance is the increase in organizational adaptivity (Teisman 2008: 358). Flexibility and changeability are more included from the start onwards in decision-making processes. At the same time, the inclusion of other stakeholders besides government is not risk-free. Accountability is a real issue when decision-making is shared (Koliba et al. 2011: 35–36; van Kersbergen and van Waarden 2004; Sørensen and Torfing 2005). Furthermore, some stakeholders are more vocal than others, which might result in an unrealistic representation of some interests above others.

Water governance—drought governance being a part of that—concerns the same multi-actor approach in the field of water. For water, governance is very important, because water is a complex and highly interconnected system which touches upon many others domains and fields such as agriculture, economic development, social development, ecology and health (Edelenbos et al. 2013: 2). This means many stakeholders are involved, each with very different stakes (Leach and Pelkey 2001; Kuks 2004). In such a field it is almost impossible to realize change in just a top-down hierarchical manner. Rather, more bottom-up, horizontal and multi-stakeholders approaches are required (Edelenbos et al. 2013: 2). Especially in the light of the fact that even though numerous methods and technologies exist to solve water problems such as water pollution, water supply and water surpluses, it is highly noteworthy that implementation is often still lacking. That has led some to argue that not a lack of water technology is what causes the current 'water crisis' (UNDP 2013: iv) but rather a lack of water governance (UNESCO 2006). Perhaps as a result of that, water governance is an upcoming theme in the field of public administration (Edelenbos et al. 2013; de Boer et al. 2013; Edelenbos and Teisman 2011; Teisman and Edelenbos 2011; Bressers and Lulofs 2010; Huitema et al. 2009; Kampa 2007; Pahl-Wostl 2007; Gopalakrishnan et al. 2005; Kuks 2004).

The rationale for the shift from government to governance is the fact that awareness has increased that monocentric government models and approaches are incapable of handling persistent uncertain and complex situation (Edelenbos et al. 2013; Kickert et al. 1997; Koppenjan and Klijn 2004; Van Buuren et al. 2010). The required adaptivity cannot be found in solely governmental steering. Koppenjan and Klijn (2004: 95–100) define governance as a fundamentally different form of response to uncertainty than traditional responses. Governance then is a way of linking complex interactions between actors in solving difficult problems (ibid.: 99), whereas more traditional responses rely more on research, go alone strategy, only limited consultation, and top-down steering.

The *degree of actor involvement* is therefore an important characteristic of the governance approach. But governance is more than 'just' including more actors. It also concerns the *multiplicity of all levels and scales* involved and the *varying problem perceptions and objectives* that occur in such a multi-scale, multi-actor environment. This reality poses an issue that has been called 'multi-level governance'. Blomquist and Schlager (1999: 7, 39–43) also emphasize the relation between the many facets of the problem and the horizontal and vertical coordination this requires. The same goes for Rosenau (2000: 10–11).

As done in this book, O'Toole (2000: 276–279) treats governance in the context of studies of the implementation of policy strategies. He adds to the multi-level, multi-actor, multi-problem perception aspects 'the multivariate character of policy action'. He refers to Milward and Provan (1999: 3), who state: 'The essence of governance is its focus on governing mechanisms—grants, contracts, agreements—that do not rest solely on the authority and sanctions of government.' *The instruments and strategies* available and required therefore also increase.

O'Toole (2000) also points to the work of Lynn et al. (2000a, b), who approach governance from the public management perspective. They begin by noting that policy programs are implemented in a web of many diverse actors. As a consequence, the model of governance they develop concentrates not only on the objectives and instruments of policy, but also the resources and organization of implementation. Their model differs from usual overviews mainly because it clearly shows that these aspects of organization and resources can take a wide variety of forms and have a multi-functional character (pp. 257–258). Peters and Pierre (1998: 226–227) also consider a/o. the 'blending of public and private resources' to be features of the governance concept. This brings a fifth element into the picture, namely that of the available *resources and responsibilities*.

A classical definition of the concept of 'policy' is that of an actor striving to attain certain goals with certain means. Compared to this concept the multiplicity of all elements is striking in the discussion on 'governance'. No longer is one dominant actor supposed to govern a certain sector, but a multiplicity of them, operating at multiple levels simultaneously influencing developments in the sector. Furthermore goals are not rationally chosen purposes, but often the result of clashes and compromise from different problem perspectives. Means not only consist of the multiplicity of policy instruments that blend in various strategies, but also of the responsibilities and resources given to again often multiple organizations to use them in practice.

Applied on implementation processes, this exploration of the governance literature leads in our opinion to the following elements of governance (Bressers and Kuks 2003):

1. Levels and scales (not necessarily administrative levels): governance assumes the general multi-level character of policy implementation.
2. Actors and networks: governance assumes the multi-actor character of policy implementation.
3. Problem perception and goal ambitions: governance assumes the multi-faceted character of the problem perceptions and resulting goal ambitions of policy implementation.

4. Strategies and instruments: governance assumes the multi-instrumental character of policy strategies for policy implementation.
5. Responsibilities and resources for implementation: governance assumes a complex multi-resource basis for policy implementation.

This leads us to define governance in general, and drought governance more specifically, as the combination of the relevant multiplicity of responsibilities and resources, instrumental strategies, goals, actor networks and scales that forms a context that, to some degree, restricts and, to some degree, enables actions and interactions. In Chap. 3 we will explain further how on the basis of this conceptualization of governance an assessment method is developed and how this methods works in practice.

1.3 A Short Overview on Existing Governance Assessment Methods and How We Relate to Them

Governance, and more specifically drought governance, thus requires an encompassing method of assessment, one that is not too strictly focused on a single aspect of the water domain. However, assessment methods for water governance are scarce, and often lack integrality or scientific foundation. As van Rijswick et al. (2014) state in their recent article on water governance assessment:

> However, an increasing amount of integral assessment approaches appear, but these approaches often lack scientific substantiation and grounding (OECD 2011, 2014). The information and knowledge base on which they rely can be very weak and fragmented. Integral and interdisciplinary assessment methods are scarce, partly for the reason that such integral and interdisciplinary assessments are particularly complex to develop and implement.

Van Rijswick et al. work on an attempt to create greater coherence between perspectives on assessment and relevant parts of water governance assessment. This results in a list of ten building blocks, varying from knowledge, values and involvement to responsibilities, regulations and arrangements (2014: 739). However, although that does create greater insight into the components of water governance and how diverse water governance assessment is, it does not yet lead to a clearly implementable method of water governance assessment.

This is the issue for most of what is out there on water governance assessment. Drought governance assessment as such is a much underdeveloped field, which is why we investigate primarily water governance as a more general field in this section. The step towards a more integral approach is increasingly made, and the relevance of water governance assessment is not widely disputed. But work that takes a step further towards concrete assessment method building is still very limited. Sometimes assessment methods take a step towards a more normative approach. An example here is the OECD Principles on Water Governance (brochure OECD 2015b). These are principles to provide a framework to understand whether water governance systems are performing optimally and help to adjust them where necessary. The analytical assessment model underlying these principles is more a building block approach with identified gaps and possible bridges between them. That is of course part of the movement towards an integrated assessment tool, but not yet a full tool itself.

In 2013 the UNDP published a report with a framework for water governance assessment. This governance assessment framework consists of three basic components (actors and institutions, governance principles and performance) (UNDP 2013: 8). Together these form 'water governance'. These components are further described, and then assessed with an eight-step method (from clarifying the objectives and conducting a stakeholder analysis through deciding an assessment framework and selecting indicators to analyzing results and communicating them) (ibid.: 18). More concretely, each of the three basic components is discussed with a 'how to' approach for its assessment. However, the discussion remains rather theoretical and general. It very accurately points at all the facets encompassing water governance and how these interact, but does not yet lead to a directly implementable assessment method.

The OECD is, however, engaged in an extensive exercise to come towards such an integrated tool, or an integrated set of indicators. As a first step they have created an overview of all indicators and assessment tools that they knew of (OECD 'Inventory of Water Governance Indicators and Measurement Frameworks'—version July 10th 2015a). In this inventory they list a whole lot of indicators, but also databases, guidelines, maps, and assessment tools. Focusing on the assessment tools, they mention 25 assessment tools, partially already listed by earlier such effort (e.g. UNDP 2013). Some of the assessment frameworks come from large supranational organizations, such as multiple assessment tools from the UN (different programs). Others include for instance the work of Van Rijswick et al. (2014) mentioned above, but also the work of the authors of this book. Many of the mentioned assessment tools have a specific focus, for instance gender (UN WWAP UNESCO, Project for Gender Sensitive Water Monitoring Assessment and Reporting), solidarity (UNDP Global water solidarity, Certificate for Decentralized Water Solidarity), or sanitation (for instance UN-Water, WHO, GLAAS Global Analysis and Assessment of Sanitation and Drinking-Water, and IDB, IWA, AquaRating). Others focus more on example setting and best practices (for instance UN-CEPAL, Best practices in regulating state-owned and municipal water utilities). What is noteworthy in this excellent overview of assessment methods is that many methods focus on specific aspects of water governance, for instance law, economy, human rights, governmental action, etc. Many instruments are also evaluation or monitoring assessment methods of specific plans, policies or actions, rather than assessment tools for a full governance setting.

The OECD inventory also includes the governance assessment method discussed in this book. For obvious reasons it is included as having a specific focus, namely drought resilience. But actually that is not entirely correct, as the basic features of the method are much wider applicable than solely for drought governance assessment. In relation to the three sections of the diagram of OECD principles (effectiveness, efficiency and engagement) we concentrate primarily on the effectiveness part of the diagram.

1.4 Towards Constructing Our Own Governance Assessment Model

Although water governance assessment is undoubtedly an upcoming theme and assessment methods are increasingly developed, we do still see a gap from the theoretical recognition that water governance assessment is needed and what components should be part of that assessment to the development of an actual hands-on but science-based assessment method. We will provide such a method in this book. This method is based on our work in an European Interreg IVb NWE-project, called Benefit of Governance in Drought Adaptation (abbreviation: DROP). In this section we will provide some insight into this background of our work before the upcoming chapters will describe our assessment method in full detail.

As we have discussed in Sect. 1.2 we view five dimensions as central to water governance assessment: (1) Levels and scales; (2) Actors and networks; (3) Problem perception and goal ambitions; (4) Strategies and instruments; (5) Responsibilities and resources. These dimensions are based on study of scientific literature and earlier research. The resulting assessment method—further discussed in upcoming chapters—has been applied on several case studies. Predecessors of the present assessment method have been used in an EU six-country study on water governance (Bressers and Kuks 2004) and a study on Greece (Kampa and Bressers 2008). Later, also further studies in The Netherlands (a/o. de Boer and Bressers 2011), Canada (de Boer 2012), Romania (Vinke-deKruijf et al. 2015) and Mexico (Franco-García et al. 2013; Casiano and Bressers 2015) were done with a further developed version of the assessment method. The final elaboration of the method and its most extensive application thus far, however, has been in the above-mentioned 'DROP project'. DROP was about drought, as a specific subfield of water management more in general. It was a project on the edge of science and practice. The project started in 2013 and continued till the end of 2015.

Eleven partners formed the project team; six regional water authorities (practice partners) and five knowledge institutions (science partners, also known as 'governance team'). The project was based in five countries: The Netherlands, Germany, Belgium, France and the United Kingdom. In each country one region—in the Netherlands two regions—was studied by the scientific partners and drought adaptation measures were implemented by the practice partner.

The practice partners in DROP implemented various drought adaptation measures. It differed per partner what was done. In the region Twente, the Netherlands, brook restoration measures were carried out, such as removing drainage systems, muting ditches, shoaling streams, and constructing water storage areas. Apart from that, water management plans were written for local farmers to aid drought adaptation on parcel level, and two studies were conducted (one about level-dependent drainage and one on surface run-off). In the region Salland, also in the Netherlands, two structures were built as part of a larger plan where the double-edged sword of too much water and too little water is addressed simultaneously by a set of structures that combine discharge and pumping functions. The project also paid attention to optimization of water management. In the region Flanders, Belgium, instruments for drought monitoring and impact modelling were set up, combined with information provision on drought. This resulted in among others the inclusion of drought as one of the four main themes on a web portal, where the developed drought indicators are published and disseminated.

In the region Eifel-Rur, Germany, the focus was on preventing deterioration of water quality in the water reservoir system. This was done by investigating possible changes in the inflow over the last decades, to see if trends could be distinguished. Based on this study, the management plans for discharge downstream of the reservoir system can be checked and if necessary adapted. In the region Brittany, France, two strands of work were carried out. The practice partner developed an innovative lock for the dam of a reservoir that prevents salt water intrusion when boats pass the dam to and from the ocean. One of the scientific partners developed a tool that forecasts inflows to the reservoir during low flow season and therefore aids anticipation of critical situations. Last, in the region Somerset, United Kingdom, a whole set of innovative approaches was implemented to increase drought resilience, examples of which are modelling and technology transfer, water demand management, soil moisture data collection and analysis, different cover crops, and all kinds of measures aimed to conserve the peat soils, such as scrub clearance, re-grading peat soils and improvement of structures that retain rainwater.

These six regions were grouped in three pilots. The pilot Nature predominantly focused on drought adaptation measures with regard to preservation of the natural environment. The two practice partners in this pilot were Twente and Somerset. The pilot Agriculture predominantly focused on drought adaptation measures in relation to agriculture. The two practice partners in this pilot were Salland and Flanders. The pilot Freshwater predominantly focused on drought adaptation measures for the preservation and management of freshwater reservoirs. The two practice partners in this pilot were Brittany and Eifel-Rur.

The scientific partners in the DROP project team, called 'the governance team', worked on governance assessment of these same six regions. They visited the regions twice, and spoke with the regional water authority, also many other regional and local stakeholders. Based on these conversations they were able to create a region diagnosis on the five dimensions we discerned above and following that diagnosis also recommendations for the future. These recommendations were multi-level; sometimes matters that could rather easily be picked up by the regional water authority itself, but in other cases also broader recommendations for the national level, with lesser possibilities to influence it directly by the regional water authority that was practice partner in the project.

The regional visits by the governance team were one type of the site visit exchanges that took place in the DROP project. Another type of exchange was the 'drought team visits'; where several experts on drought from the water authority went to visit their partner region; for instance Twente experts visited the region Somerset to learn about drought adaptation for nature there. Likewise, experts from Somerset visited Twente. These drought team exchanges took place twice, just like the governance team visits. A third set of exchange was the stakeholder exchange. In this exchange a group of approx. 5–10 stakeholders visited the partner region. These stakeholders were representatives from local/regional governmental agencies, NGOs, businesses also local farmers. These visits took place once during the project's duration. A last form of exchange was the full partnership meetings of the project team, where partners shared their work thus far, the lessons they had learned, and the plans they had for the future.

Exchange and mutual learning were therefore important aspects of the work in DROP. For governance assessment this meant that a strong focus laid on interaction and exchange, and that there was a lot of room for discussion, and on the spot science. This gives our assessment model a specific place in the wider array of assessment approaches out there, as discussed in Sect. 1.3. Distinguishing characteristic of our approach—compared to other governance assessment approaches —are the following ones. (1) Many approaches of governance mix elements of descriptive nature and elements of normative nature, while our approach tries to clearly separate the descriptive elements (the five dimensions of governance discussed at the end of Sect. 1.2) and the normative aspects (four criteria we employ in our assessment, namely intensity, flexibility, extent and coherence). (2) Furthermore our approach derives the normative criteria from a specific goal, namely the feasibility and likelihood of realization of a certain category of measures or projects (in this case the promotion of drought resilience). Thus the normative component is limited and focused. That does not mean that more ethical approaches ('good governance' like the one of the OECD) are wrong, just that they have another focus. Our approach could be considered more practice-oriented—with the risk that it can only be applied ethically in cases that the projected policies and projects serve 'good' goals. (3) Our approach makes a clear separation between the conditions and the activities. In many approaches 'governance' is used for both the process and the contextual conditions for the process. In our approach 'governance' is just used for the context. It is even a very distinguished characteristic of our approach that not everything (the circumstances and the process) is put in this one basket, but that the governance context and the process are seen as related but separated so that it is possible to study the impact of the governance conditions on the process. This again makes it relatively practice-oriented.

1.5 Outlook and Reader Guidance

In this book the work conducted in the DROP project is discussed and complemented with additional comparative analysis. We have constructed the book in several distinctive sections. The first section of the book provides an introduction to our work. This chapter is the first step in that.

Chapter 2 (Stein et al.) elaborates on the European policy perspective on drought and water scarcity. Stein et al. provide the reader with extensive knowledge into the directives and plans behind current European perspectives on drought. In doing so they show how the past two decades have seen a transition from scattered policies on generally broader water governance issues towards more direct policy actions to adapt to and mitigate droughts. Despite more European attention, the effectiveness of drought policies still largely depends on the national and regional translations into initiatives and plans.

As a result of that the study of the national and regional governance context becomes all the more important in assessing drought resilience. Chapter 3 (Bressers et al.) discusses in detail the Governance Assessment Tool as developed and refined in the DROP project. The authors discuss the origins of the tool in contextual interaction theory, the dimensions and criteria that form the backbone of the tool and the matrix that originates from these dimensions and criteria. In these matrix evaluative questions are formulated that can be asked to local and regional stakeholders. Based on their answers and insights a judgment can be reached on whether the governance circumstances investigated in that matrix box are supportive, restrictive or neutral for drought adaptation. Through the collection of data on all matrix boxes a visualization can be developed which shows in one quick glance the governance state of affairs in that region. To create a more precise visualization, arrows can be added to each box indicating upward or downward trends for that box. This inserts a longitudinal aspect into the visualization. Chapter 3 ends with a discussion of the GAT application in the DROP project, in order to discuss our lessons learned and problems and opportunities we ran into while applying the instrument.

After this chapter we precede to the second section of the book; that of the case chapters. Each of the six regions from the DROP project is discussed as a case study here, and chapter authors show how they have applied the GAT to that case, what results they found, and what main messages they distill from that.

Chapter 4 (Vidaurre et al.) discusses case study region Eifel-Rur. Based on the application of the GAT they conclude the current governance situation in Eifel-Rur is 'intermediate', hovering between fully supportive or fully restrictive. Especially the flexibility (room for manoeuvering) and the intensity (sense of urgency) of drought governance in Eifel-Rur have much room for improvement. At the same time, the authors witness an already occurring improvement in these criteria. As a result of their analysis the authors reach a list of recommendations for the Eifel-Rur region. For example, they advise to diversify strategies for drought preparedness by connecting with water scarcity and climate change debates, to further develop water demand management, and to increase synergy with farmers.

In chap. 5 (Browne et al.) the region Somerset is analyzed. In Somerset there was a shift from increasing drought awareness when the project started to resistance to the topic of drought during later phases of the project, as a result of severe flooding in the winter of 2013–2014. The chapter authors discuss the fragmented nature of the English water sector and the split responsibilities that exist as a result of that. The discussion on water and drought in the UK is very politicized and emotive, and this has an impact on how the topic has to be addressed in order to increase drought resilience. To deal with this, the chapter authors call upon the decision-makers in the UK to engage in collaborative processes of water governance instead of the current silos.

Chapter 6 (La Jeunesse et al.) assesses the French region Brittany, and therein the Vilaine river basin. The governance of the Vilaine river basin, and more specifically of the area around the Arzal dam, largely revolves around the multi-user conflict in the area and realizing drought awareness. Awareness about the effects of climate change on drought is low. The general judgment of the region's governance is moderately positive, but due to the limited attention for climate change and insufficient knowledge about the effects of drought on the area, the authors advise to enhance knowledge and cooperation on climate change and its impacts on drought, and increase foresight and sharing forecasting information.

The next chap. 7 (Troeltzsch et al.) gives an account of the governance situation in the Belgian region Flanders. The authors classify the current state of affairs as 'intermediate'. Especially in terms of responsibilities and resources there is room for improvement. A reason for this is the fact that there is no assigned budget for drought. Overall, the authors say that Flanders is at early stage of establishing drought resilience measures. Through first activities motivation is increasing, and increasingly water scarcity and drought are integrated in some general water management strategic documents. The authors discuss how further realization can be sought through increasing awareness, mainstreaming drought risks and engaging with other public actors.

Chapter 8 (Özerol et al.) presents the results for the Dutch region Salland. They write that Salland has a neutral governance context regarding its drought resilience policies and measures. For Salland the most supportive dimension is levels and scales, whereas the coherence of strategies and instruments and the intensity of problem perspectives and goal ambitions are the restrictive contextual factors. Most matrix boxes are scored neutral. Important explanations of this are that drought measures are not integrated into existing water use and the dominance of flood management over drought management. Positive aspects are the existence of trust and collaboration. As a result, the authors recommend to increase awareness and understanding of drought (management) and actively enable non-governmental parties to share responsibilities.

Chapter 9 (Bressers et al.) discusses the last case study; the Dutch region Twente. The governance assessment results of this region are mixed, leading to a moderately positive general judgment; varying from excellent in one box to restrictive for other boxes. Excellent is the actor coherence in Twente, but more restrictive are for example the slow integration of the drought resilience awareness and the resulting reliance on voluntary preventive measures. Just like other chapters the recommendations for the region Twente partly also concern matters such as awareness and exchange. Another, more Twente-specific, recommendation is the upscaling from farm-level approach to full area-level approach, where work transcends farm-level voluntary measures (although those should be continued as well) and also includes larger scale measures to create more synergy between participating actors.

The third section consists of cross-cutting perspective chapters on the three topics of Nature, Agriculture and Freshwater—consistent with the three pilots of DROP.

Chapter 10 (Özerol and Troeltzsch) is the first of these cross-cutting chapters, and discusses the topic Agriculture. The chapter shows how there is a tension between the fact that agriculture is a key water user, therefore significantly impacting drought and water scarcity circumstances, yet at the same time not being prioritized over for instance drinking water and energy production. Awareness, both public and private, is low, also due to the low visibility of drought in Northwest Europe. Increasing this awareness has much potential for improving the way drought and water scarcity are tackled in agricultural production.

Chapter 11 (Furusho et al.) discusses the topic of Freshwater. Here, the same low visibility of drought hampers the uptake of drought adaptation measures. However, the authors point to the fact that everyone agrees with the importance of freshwater availability, and hence the topic of safeguarding future freshwater availability can be used as an entryway into more drought awareness. In order to facilitate this awareness building process the authors plead for more monitoring of water withdraws. The greater insight in the effects of water shortages in freshwater production can help to trigger action.

The last cross-cutting chapter, chap. 12 (Bressers and Stein) discussed the field of Nature. They apply contextual interaction theory to discuss the main conclusions for this field. Motivation is highly varied, and interestingly for cognition the authors conclude that awareness that drought is becoming a topic of increasing importance is recognized widely by nature conservation actors. It appears that for Nature the actors involved recognize the importance of drought better than for Agriculture or Freshwater. Unfortunately, these same nature conservation actors have limited resources. This means that for nature not the awareness among primary stakeholders themselves is the biggest problem, but their possibilities to address the issue properly are.

In chap. 13 (Larrue et al.) we discuss our application of the Governance Assessment Tool and the generic recommendations we can draw from the case studies and cross-cutting studies. The aim of this book as such is a double focus on both model development and refinement, as well as real-life application to regional drought adaptation. In line with the observations in the outlook above the chapter concludes with four overarching conclusions for the whole book: (1) Continuous focus on realizing awareness is needed, (2) An increase in preparation and implementation of water demand management is required, (3) Flood and drought management need to be integrally dealt with, and (4) Tailored action is key in tackling drought and water scarcity effectively due to regional diversity.

References

Blomquist W, Schlager E (1999) Watershed management from the ground up: political science and the explanation of regional governance arrangements. Paper presented at annual meeting of the american political science association, Atlanta, 2–5 Sept 1999

Bressers N (2011) Co-creating innovation. A systemic learning evaluation of knowledge and innovation programmes. Dissertation, Erasmus University of Rotterdam

Bressers H, Kuks S (2003) What does "governance" mean? From conception to elaboration. In: Bressers H, Rosenbaum W (eds) Achieving sustainable development: the challenge of governance across social scales. Praeger, Westport Connecticut, pp 65–88

Bressers H, Kuks S (eds) (2004) Integrated governance and water basin management. Conditions for regime change and sustainability. Kluwer Academic Publishers, Dordrecht

Bressers H, Lulofs K (eds) (2010) Governance and complexity in water management. Creating cooperation through boundary spanning strategies. Edward Elgar, Cheltenham

Casiano C, Bressers H (2015) Changes without changes. The Puebla's Alto Atoyac sub-basin case in Mexico. Water Gov 1(2):12–16

COPA COGECA (2003) Assessment of the impact of the heat wave and drought of the summer 2003 on agriculture and forestry

de Boer C (2012) Contextual water management. A study of governance and implementation processes in local stream restoration projects. University of Twente, Enschede

de Boer C, Bressers H (2011) Complex and dynamic implementation processes. Analyzing the renaturalization of the Dutch Regge river. University of Twente and Water Governance Centre, Enschede, The Hague

de Boer C, Vinke-de Kruijf J, Özerol G, Bressers HTA (eds) (2013) Water governance, policy and knowledge transfer. International studies on contextual water management. Earthscan—Routledge, Oxford

EC (2007) Communication from the Commission to the European Parliament and the Council—addressing the challenge of water scarcity and droughts in the European Union. COM (2007) 414 final, European Commission, Brussels. http://eur-lex.europa.eu/legal-content/EN/TXT/PDF/?uri=CELEX:52007DC0414&from=EN. Accessed 14 Dec 2015

Edelenbos J, Teisman GR (2011) Symposium on water governance. Prologue: water governance as a government's actions between the reality of fragmentation and the need for integration. Int Rev Admin Sci 77(1):5–30

Edelenbos J, Bressers N, Scholten P (2013) Water governance as connective capacity. Ashgate Publishing, Farnham

Franco-García L, Hendrawati-Tan L, Gutiérrez-Díaz EC, Casiano C, Bressers H (2013) Institutional innovation of water governance in Mexico. The case of Guadalupe Basin, near Mexico City. In: de Boer C, Vinke-de Kruijf J, Özerol G, Bressers HTA (eds) (2013) Water Governance, Policy and knowledge transfer. International studies on contextual water management. Earthscan—Routledge, Oxford, UK, New York, USA, pp 188–204

Gopalakrishnan C, Tortajada C, Biswas AK (eds) (2005) Water institutions. Policies, performance and prospects. Springer, Berlin

Huitema D et al (2009) Adaptive water governance. Assessing the institutional prescriptions of adaptive (Co-)management from a governance perspective and defining a research agenda. Ecol Soc 14(1) art 26

Kampa E (2007) Integrated institutional water regimes. Realisation in Greece. Logos, Berlin

Kampa E, Bressers H (2008) Evolution of the Greek national regime for water resources. Water Policy 10(5):481–500. ISSN 1366-7017

Kickert WJM, Klijn E-H, Koppenjan JFM (eds) (1997) Managing complex networks. Strategies for the public sector. SAGE Publications, London

Klijn E-H, Koppenjan JFM (2000) Public management and policy networks: foundations of a network approach to governance. Pub Manage 2(2):135–158

Koliba C, Meek JW, Zia A (2011) Governance networks in public administration and public policy. CRC Press, Boca Raton

Kooiman J (1993) Findings speculations and recommendations. In: Kooiman J (ed) Modern governance: new government—society interactions. SAGE, London, pp 249–262

Kooiman J (2003) Governing as governance. Part III. SAGE Publications Ltd, London

Koppenjan J, Klijn E-H (2004) Managing uncertainties in networks. A network approach to problem solving and decision making. Routledge, London

Kuks SMM (2004) Water governance and institutional change. Dissertation. University of Twente, Enschede

Leach WD, Pelkey NW (2001) Making watershed partnerships work. A review of the empirical literature. J Water Resour Plann Manage 127(6):378–385

Lynn LE Jr, Heinrich CJ, Hill CJ (2000a) The empirical study of governance: theories models and methods. Georgetown University Press, Washington DC

Lynn LE Jr, Heinrich CJ, Hill CJ (2000b) Studying governance and public management: challenges and prospects. J Pub Adm Theor 10:233–261

Milward HB, Provan KG (1999) How networks are governed (unpublished paper)

O'Toole Jr LJ (2000) Research on policy implementation. Assessment and prospects. J Pub Adm Theor 10:263–288

OECD (2011) Water governance in OECD countries. A multi-level approach. OECD publishing, Paris

OECD (2014) Water governance in the Netherlands. Fit for the future? OECD, Paris

OECD (2015a) OECD inventory. Water governance indicators and measurement frameworks. http://www.oecd.org/gov/regional-policy/Inventory_Indicators.pdf

OECD (2015b) OECD principles on water governance. http://www.oecd.org/gov/regional-policy/OECD-Principles-on-Water-Governance-brochure.pdf. Accessed 29 Sept 2015

Pahl-Wostl C (2007) Transitions towards adaptive management of water facing climate and global change. Water Resour Manag 21(1):49–62

Pereira L, Cordery I, Iacovides I (2002) Coping with water scarcity, IHP-VI, Technical Documents in Hydrology, 58, UNESCO, Paris. http://unesdoc.unesco.org/images/0012/001278/127846e.pdf. Accessed 22 Oct 2015

Peters BG, Pierre J (1998) Governance without government? Rethinking public administration. J Pub Adm Theor 18:223–243

Rosenau JN (2000) The governance of fragmentation: neither a world republic nor a global interstate system. Paper presented at IPSA world conference, Quebec

Sørensen E, Torfing J (2005) The democratic anchorage of governance networks. Scand Polit Stud 28(3):195–218

Teisman GR (2008) Complexity and management of improvement programmes. Pub Manage Rev 10(3):341–359

Teisman GR, Edelenbos J (2011) Towards a perspective of system synchronization in water governance. A synthesis of empirical lessons and complexity theories. Int Rev Adm Sci 77 (1):101–118

Termeer CJAM (2009) Barriers to new modes of horizontal governance. A sense-making perspective. Pub Manage Rev 11(3):299–316

UNDP (2013) User's guide on assessing water governance. http://www.undp.org/content/undp/en/home/librarypage/democratic-governance/oslo_governance_centre/user-s-guide-on-assessing-water-governance.html. Accessed 16 Dec 2015

UNESCO (2006) Water, a shared responsibility. UNESCO/Berghahn Books, Barcelona

Van Buuren A, Edelenbos J, Klijn E-H (2010) Gebiedsontwikkeling in Woelig Water. Over water governance bewegend tussen adaptief waterbeheer en ruimtelijke besluitvorming. Boom Lemma uitgevers, Den Haag

van Kersbergen K, van Waarden F (2004) Governance' as bridge between disciplines. Cross-disciplinary inspiration regarding shifts in governance and problems of governability, accountability and legitimacy. Eur J Polit Res 43:143–171

van Rijswick M, Edelenbos J, Hellegers P, Kok M, Kuks S (2014) Ten building blocks for sustainable water governance—an integrated method to assess the governance of water. Water Int 39(5):725–742

van Schie N (2010) Co-valuation of water. An institutional perspective on valuation in spatial water management. Dissertation, Erasmus University of Rotterdam

Vinke-deKruijf J, Kuks SMM, Augustijn DCM (2015) Governance in support of integrated flood risk management? The case of Romania. Environ Dev 16:104–118

Wilhite DA, Glantz MH (1987) Understanding the drought phenomena: the role of definitions. In: Wilhite DA, Easterling WE (eds) Planning for drought. Towards a reduction of societal vulnerability. Westview Press, Boulder

Chapter 2
European Drought and Water Scarcity Policies

Ulf Stein, Gül Özerol, Jenny Tröltzsch, Ruta Landgrebe, Anna Szendrenyi and Rodrigo Vidaurre

2.1 Introduction: Drought Events and the Importance of Policy Responses on the European Level

Over the last decade, Europe's drought management and policy has been characterized by a predominantly crisis-oriented approach. However, the widening gap between the impacts of drought episodes and the ability to prepare, manage and mitigate such droughts has motivated the European Union (EU) to make significant improvements that address drought management using a preventative approach (Kampragou et al. 2011). Not surprisingly, disaster response and recovery policy, disaster prevention, and mitigation and preparedness approaches have become increasingly more widespread.

That said, in order to tackle drought risk and its impacts, an integrated approach to water governance is needed, one that considers multiple dimensions of water management (Bressers et al. 2013). Such an increased demand for more sustainable and proactive policies must stem from all sectors, including agriculture, urban development, energy, nature conservation, and recreation. To create a drought-resilient society, equipped with the appropriate tools and abilities to respond and cope with the impacts of extreme events such as droughts, requires development of long-term strategies and processes to address and reduce the risks of drought (Kampragou et al. 2011).

This chapter focuses on the main EU policies related to drought and water scarcity and highlights recent policy developments in all relevant sectors. Additionally, it provides an overview of those European policies that impact drought and

U. Stein (✉) · J. Tröltzsch · R. Landgrebe · A. Szendrenyi · R. Vidaurre
Ecologic Institute, Pfalzburger Strasse 43/44, 10717 Berlin, Germany
e-mail: ulf.stein@ecologic.eu

G. Özerol
CSTM - Department of Technology and Governance for Sustainability, University of Twente, 217, 7500 AE Enschede, The Netherlands
e-mail: g.ozerol@utwente.nl

© The Author(s) 2016
H. Bressers et al. (eds.), *Governance for Drought Resilience*,
DOI 10.1007/978-3-319-29671-5_2

drought-related management issues, through an examination of legal, organizational, financial and political issues that guide and structure the interactions among all actors.

Over the past thirty years, there has been an increasing trend in droughts events and their impacts in the EU. Water is relatively abundant in much of Europe; however, large areas are affected by water scarcity and droughts (Kazmierczyk et al. 2010). Water scarcity affects at least 11 % of the European population and 17 % of EU territory (European Union 2010); it is experienced by various member states and not limited to the Mediterranean region (European Commission 2011). The comparison of the periods 1976–1990 and 1991–2006, that shows a doubling in both area and population affected (European Environmental Agency 2010) and the quadrupled yearly average costs (European Union 2010). One of the worst droughts occurred in 2003, when one-third of EU territory and over 100 million people were affected (European Union 2010); see also the Box 1.1 in Fig. 2.1). The State of the Environment Report states that "except in some northern and sparsely-populated countries that possess abundant resources, water scarcity occurs in many areas of Europe, particularly in the south, confronted with a crucial combination of a severe lack of and high demand for water" (Kazmierczyk et al. 2010).

Climate change is further projected to increase water shortages across Europe. The most severe impacts are expected in southern and southeastern regions, which already suffer from water scarcity. These areas will face reductions in water availability as more frequent and intense drought events occur. While water availability will generally increase in northern regions, in summer periods availability of water may decrease and lead to drought spells (Jol et al. 2008 in European Environmental Agency 2010).

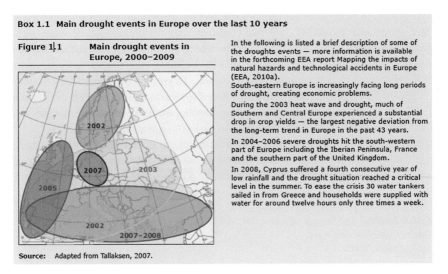

Box 1.1 Main drought events in Europe over the last 10 years

Figure 1.1   Main drought events in
                 Europe, 2000–2009

In the following is listed a brief description of some of the droughts events — more information is available in the forthcoming EEA report Mapping the impacts of natural hazards and technological accidents in Europe (EEA, 2010a).

South-eastern Europe is increasingly facing long periods of drought, creating economic problems.

During the 2003 heat wave and drought, much of Southern and Central Europe experienced a substantial drop in crop yields — the largest negative deviation from the long-term trend in Europe in the past 43 years.

In 2004–2006 severe droughts hit the south-western part of Europe including the Iberian Peninsula, France and the southern part of the United Kingdom.

In 2008, Cyprus suffered a fourth consecutive year of low rainfall and the drought situation reached a critical level in the summer. To ease the crisis 30 water tankers sailed in from Greece and households were supplied with water for around twelve hours only three times a week.

Source: Adapted from Tallaksen, 2007.

Fig. 2.1 Recorded drought events in Europe between 2000–2009. As shown, the year 2003 drought disproportionately impacted much of South and Central Europe (Tallaksen 2007 in European Environmental Agency 2010). *Source*: http://www.eea.europa.eu/legal/copyright

Issued in 2000, the Water Framework Directive (WFD) established the EU wide framework for water management, incorporating tools to achieve 'good status' of all European waters (European Community 2000). The quantitative water issues with regards to water scarcity and droughts were identified as a gap in the implementation of the WFD, which influence the achievement of the environmental objectives, and therefore the good status of waters. The occurrence of major drought events during this period (in particular in 2003) further increased these concerns. They were captured by the EU Water Directors during the informal meeting in Rome in 2003, where the development of an initiative on water scarcity issues was agreed upon, and a technical document on drought management and long-term imbalances was consequently prepared and presented to the Water Directors Meeting in 2006. The document showed that these phenomena have been increasing in intensity and extent at European level in the last decades, with worsening socio-economic and environmental impacts. Therefore, in the same year (2006), some member states requested European action on water scarcity and drought events at the Environment Council, raising concerns on the need for further development at political and technical levels to address the environmental, social and economic impacts of water scarcity and drought (WS&D) (Informal Council of Environment Ministers 2007). In response the Environmental Council requested for actions on this issue in 2006 and within the Common Implementation Strategy (CIS) of the WFD, the European Commission has conducted several analyses of WS&D in the EU (i.e. (European Commission—DG ENV 2007). This assessment showed that water scarcity and drought events affect all EU countries in a variety of ways (Informal Council of Environment Ministers 2007) and now have an added European dimension, as drought and water scarcity is no longer exclusive to southern European countries (Portuguese Ministry of Environment, Spatial Planning and Regional Development 2007). This shift provided the basis for establishing a common approach at European Union level. The EU Presidency of Portugal (2007) placed water scarcity and droughts as one of its main environmental policy priorities. It welcomed the Communication 414 final (European Commission 2007b) on 'Addressing the challenge of water scarcity and droughts in the European Union' and succeeded in making WS&D an active environmental policy area with specific instruments and strategies (Portuguese Ministry of Environment, Spatial Planning and Regional Development 2007). The Communication summarized the main trends and concerns and identified a series of actions to be taken at EU and national level, giving priority to water savings and water efficiency measures, and further integrating water issues into all cross-cutting policies. It also emphasized the importance of taking stock of climate change and adaptation policy areas, such as agriculture (Farmer et al. 2008; European Environmental Agency 2010).

The Environment Council was supportive of the Communication and invited the Commission to review and further develop the evolving EU strategy for WS&D by

2012 (European Commission—DG ENV 2012). In 2008, the European Parliament adopted a report on the Communication, calling for a number of initiatives at the EU level. It also requested the Commission to initiate pilot projects in areas of key interest (European Commission—DG ENV 2012).[1]

In January 2008, MEPs and experts convened at the European Parliament's Climate Change Temporary Committee's Fourth Thematic Session to discuss the complex links between water issues and climate change. The session determined that significant decisions regarding the best way forward with climate adaptation strategies was a top priority. Specifically, the discussion focused on the need for water policies to respond to the impacts of climate and changing agricultural demands, calling policy development to move beyond water policy itself to be more encompassing (Farmer et al. 2008).

The former European Parliament Environment Commissioner Potočnik was a key leader in mobilisations efforts following the session. At a hearing in January 2010, he announced a new focus area on resource efficiency, including water efficiency, for the upcoming mandate. He also unveiled a new Commission initiative entitled "Blueprint for safeguarding European Waters," slated for release in 2012. The Blueprint aimed to review the WFD, including the successes and challenges of its implementation, as well as provide insight on water and resource vulnerabilities in the face the climate change and other man-made pressure (European Commission—DG ENV 2012).

The European Commission Joint Research Centre also helped to establish the European Drought Observatory (EDO) as part of ongoing efforts to integrate drought into policy. Since 2011, the EDO has been the leading disseminator on drought-relevant information and maps of indicators derived from a range of different data sources, including precipitation measurements, satellite measurements and modelled soil moisture content (Vogt et al. 2011). At its core, the EDO serves to buttress the already existing Global Drought Information System, with a focus on the European context.

Like its sister initiative, the EDO provides a technical approach to drought policy and integrates "relevant data and research results, drought monitoring, detection and forecasting on different spatial scales, from local and regional activities to a continental overview at EU level (Vogt 2011)" in order to aid evaluation and decision-making of future water scarcity and drought events (Vogt et al. 2011). The EDO is also responsible for severe drought events, and produces reports detailing the situation to better inform policymakers (European Commission—JRC 2015). In addition, the EDO is also responsible for retrieving information on droughts and related topics from global news portals using the European Media Monitor tool (Council of the European Communities 1979).

---

[1]See also: http://ec.europa.eu/environment/water/quantity/scarcity_en.htm.

2.2 Policy Frameworks for the European Governance Structure

Understanding the policy framework is essential for analyzing the governance structures in general. All five dimensions of governance (cp. Chaps. 1 and 3) are directly or indirectly linked to the respective policy framework. Developed over time from a series of scattered policies, the overarching drought and water scarcity policy approach in the EU is undeniably complex. To make sense of this complexity, an applied framework for understanding the mix of policies is necessary. This section introduces and describes a typology of tools for unpacking this complexity, known as the policy mix concept (Flanagan et al. 2011). Later in the chapter, we will explain the dimensions of the policy mix concept in more detail. We will also apply this comprehensive policy mix concept as a conceptual tool for deeper analysis of each policy we discuss.

2.2.1 Drought Policy Context

Concepts within the broad arena of water scarcity and drought are often not clearly differentiated. This may hamper effective implementation of policies and measures, as the lack of clear definitions and methods for analysis and adaptation cannot adequately address the inherent drivers and pressures (Schmidt et al. 2012; Vanneuville et al. 2012). Thus, in order to delineate adequate responses, defining both drought and water scarcity is essential (see also Chap. 1).

As briefly touched upon in previous sections, there are three types of drought policy responses: the post-impact (often crisis-oriented) drought policy approach, the pre-impact drought policy approach (often vulnerability reduction and resilient oriented), and the development and implementation of preparedness plans and policies (often focused on institutional capacity, including organizational frameworks and operational arrangements) (Wilhite et al. 2014).

A harmonized approach to drought risk management is still lacking both at the EU level and at the member state level. Consequently, the regional level also lacks the full integration of drought risk management into relevant water policies (Kampragou et al. 2011).

A core question for water scarcity and drought policy to consider centres around supply and demand: Should the government respond to growing water uses by finding additional supply to increased demands, or should it implement measures to curb water use and encourage efficiency (European Environmental Agency 2009; Water Scarcity and Droughts Expert Network 2007).

From the supply-side approach, policy measures often encourage restoration and improvement of existing water infrastructures and/or continued usage and expansion of natural catchments and aquifers. The demand-side approach, on the other hand, promotes policy measures that encourage subsidies and water efficiency strategies, such as reducing leaks infrastructure, smarter water use for agricultural purposes, public and water user education on conservation, as well as tailored pricing schemes and policies.

To address this central question, the (European Commission 2007a, 2010) formulated three policy options: the first, Option A, takes a supply approach; the second, Option B, aims for a water pricing approach; and the third, Option C, offers an integrated approach based on water efficiency. The three options are elaborated in greater detail in Table 2.1.

Option B is closely associated with so-called 'economic policy instruments' (EPIs). EPIs are believed to play a foundational role in shaping and achieving WS&D policy goals in the future (Mysiak and Maziotis 2012). In this context, EPIs are designed to foster efficient allocation and use of water, and cover a range of different instruments, including pricing, trading and risk sharing. Already, EPIs have contributed to making provision of water service financially sustainable by converting payments on the use of water into working incentives for water conservation. Moving forward, EPIs have the real potential to promote individual actions to save water, increase water efficiency, improve water quality and reduce water-associated risks. Thus EPI are also an important building block of Option C.

For Option C it is important to know that additional water supply infrastructures will be considered only when all other options have been exhausted, with priority for effective water pricing policy and cost-effective alternatives. Embedded within policy options set forth, the European Commission also makes it clear that water uses should be prioritized, with overriding priority to public water to ensure access to adequate water provision.

As highlighted in Table 2.1, to supplement the policy options offered by the Commission, the EC provides additional policy instruments that work towards: (1) putting the right price tag on water, (2) allocating water and water-related funding more efficiently, (3) improving drought risk management and (4) considering additional water supply infrastructures (European Commission 2007a). The European Commission suggests specific actions to this end, such as improving land

Table 2.1 Options for addressing supply, demand, and integrated approach to water scarcity and drought in Europe (European Commission 2007a, 2010)

Policy option		Actions
A	Water supply only	• Enhance development of new water supply based on existing EU legislation • Support widespread development of new water supplies, with priority to EU and national funds
B	Water pricing policies only	• Effective water pricing • Cost recovery
C	Integrated approach	• Support efficient water allocation and sustainable land use planning • Foster water efficiency technologies and practices • Foster emergence of a water-saving culture in Europe • Provide new water supply

use planning, financing water efficiency, developing drought risk management plans and early warning system on droughts, and further optimizing the use of EU Solidarity Fund and European mechanism for Civil Protection are suggested actions by the European Commission. Across all policy instruments and actions, improving knowledge and data collection regarding water quantity is a critical first step.

Despite such suggested policy options and supplemental actions, currently, the European Commission notes that most measures applied by the member states target pressures, status and impacts, and lack focus on targeting key drivers (European Commission 2012a, c). Adoption of policy instrument mixes, which include water conservation, agricultural stewardship and awareness-raising campaigns, are highly recommended to combat this gap (EEA 2009; Water Scarcity and Droughts Expert Network 2007).

2.2.2 EU Drought Policy Objectives

At the EU level, drought policy objectives share common themes. They include: (1) promoting risk management policies, (2) promoting drought preparedness and mitigation and planning measures and (3) consideration of financial assistance tools (Wilhite et al. 2014).

Engaging risk assessment and addressing management practices are essential to drought policy objectives moving forward (Kampragou et al. 2011). A risk management approach seeks to address hazard prediction and vulnerability, centering on pre-disaster preparedness measures and long-term risk reduction as means to reduce vulnerability and increase drought resiliency (Kampragou et al. 2011).

Such an approach is best captured by drought preparedness policy, which refers to actions undertaken prior to drought events intentionally designed to improve the ability of institutions to appropriately respond to a drought event operationally. This is most often accomplished through drought mitigation, which refers to actions undertaken prior to drought events designed to minimize impacts on people, the economy and the environment (Kampragou et al. 2011). Even though drought events are highly variable and geographically specific, differing in intensity, duration and spatial extent, general guidelines for processes and measures to be applied and implemented in the event of a drought are essential (Kampragou et al. 2011).

2.2.3 Policy Instrument, Measures and Strategies

In the literature, the definition of policy instruments is diverse and widely debated. In this book, **policy instruments** are defined as the fluid tools, techniques or mechanisms for achieving overarching policy objectives, in this case: the establishment of drought resilience (Bressers and O'Toole 2005; Flanagan et al. 2011;

Reichardt and Rogge 2015). Specifically, here we consider policy instruments that diffuse goal-oriented influence (also known as an intervention) of one or more actors that in turn produce effects over entire populations or very large target groups (Kaufmann-Hayoz et al. 2001), primarily in the public sector, over time. In the process of this influence, the behaviour of the target population is transformed in a structured way.

Generally, the following policy instruments types can be distinguished (Kaufmann-Hayoz et al. 2001):

- Regulatory (Command and control) instruments (e.g. water licenses)
- Economic instruments (e.g. water abstraction charges, compensation for crop losses)
- Infrastructure (Service) instruments (e.g. co-financing of water-saving infrastructure by the means of Rural Development Policy (DG AGRI), Structural and Cohesion Funds (DG REGIO), LIFE + Funds (DG ENV))
- Collaborative instruments (e.g. CGIAR Fund, ACP-EU Water Facility)
- Information (Communication) instruments, e.g. the European Drought Observatory

Policy instruments are not to be confused with **policy measures**, though they are sometimes used interchangeably. Policy measures indicate the concrete realization of a policy instrument and represent the tangible means for achieving objectives determined in the formulation of the policy instrument. To this end, policy measures serve to validate the policy instrument.

Policy strategies, on the other hand, denote the strategic orientation and management of policy instruments and their policy measures within a policy mix, in order to achieve the vision put forth by policy objective. It refers to both the ends and the means, and thus the policy objectives and principal plans of a policy strategy are closely interlinked through the proposed policy strategy (Kaufmann-Hayoz et al. 2001).

Finally, a **policy mix,** for the purposes of this chapter, is defined as the complex arrangement of policy instruments, measures and strategies that interact via dynamic processes to influence and achieve a specific broader objective. According to Reichardt and Rogge (2015) a policy mix is shaped by three defining building blocks: (1) inherent to policy mixes are the consideration of their *complexity and dynamism*; (2) within policy mixes, there is a need for identification and integration of relevant *policy processes*; (3) the incorporation of a *strategic component*. Central to the policy mix concept is the element of *policy interactions*, or the interplay among actors, instruments, measures and strategies, which operate in a multi-level and multi-actor context (Flanagan et al. 2011). This framework serves as a starting point for building up more sophisticated conceptualizations of the policy mix in the drought context, as explored in the following sections.

2.3 European Drought Policy: Policy Relations Between Flooding, Drought, Agriculture and Nature

The structure of the remainder of the chapter is organized so that relevant policy instruments on water scarcity and drought at the EU-level are explored holistically. Because there is a wide entry point for discussing water scarcity and drought-related policy instruments, we chose to focus on three perspectives, namely: nature (or conservation-based perspectives), water (specifically, the water management perspective) and agriculture (including the land management perspective). We recognize that other key perspectives exist, including the land planning perspective and the socio-economic perspective, among others.

In addition, as describing individual policy instruments would simply produce a long list and dilute the purpose of this chapter, a more systems approach is applied in order to explore relevant water scarcity and drought policy mixes. Such an approach inherently introduces more complexity. At the same time, it aims to distill that complexity to produce a comprehensive understanding of the policy landscape at hand.

In consequence, directives, often composed of several interacting policy instruments, are explored alongside Communications, often a single policy instrument interconnected with and embedded into larger the broader policy space. The aim of the section in suite is to provide an entry point for untangling the policy mixes and the relationships and interactions among the individual policy instruments. As a result, each section applies the categories developed by Landgrebe et al. (2011): (1) history, aims and objectives, (2) structure, and components and implementation and (3) relevance to drought policy implementation in order to structure the analysis.

In the following sections, we examine eight policy mixes: the EU Climate Adaptation Policy, the European Commission's Communication 'Blueprint to Safeguard Europe's Water Resources', the EU Water Framework Directive, the EU Floods Directive, the EU Habitat and Birds Directives, the EU Groundwater Directive, and the European Common Agriculture Policy (CAP). As already touched on, each policy mix applies one or more of the three perspectives outlined above as an entry point for analysis.

Lastly, it is also important to note that this chapter acknowledges that water scarcity and drought policies at the EU level also operate horizontally, such as within the Environmental Impact Assessment (EIA) and Strategic Environmental Assessment (SEA) Directives, the Sustainable Development Strategy (European Council DOC 10917/06), (Council of the European Communities 1979), and cohesion policies (Cohesion Funds and Structural Funds). However, due to the scale of these regimes, they remain outside the scope of this chapter.

EU Climate Adaptation policy is one of the main drivers for activities related to WS&D. This policy is aiming at reducing the vulnerability of relevant sectors (e.g. agriculture, tourism, industry, energy and transport) and thus mainstreaming of climate change aspects into other EU policies is the main priority.

The Review of the European Water Scarcity and Droughts Policy emphasizes that "climate change is expected to worsen the impacts of already existing stresses on water as changes in precipitation, combined with rising temperatures, will cause significant changes in the quality and availability of water resources". Therefore, the policy responses to water scarcity and drought should include adaptation measures (European Commission 2012c). According to the results of the ClimWatAdapt project (Flörke et al. 2011)—changes in water withdrawals will be the main driver of the changes in future water scarcity.

2.3.1 EC Communication on Water Scarcity and Drought

The Europeans Commission's Communication on water scarcity and drought highlights the need for increased integration of WS&D policy and policy objectives into existing policy frameworks. Though the Communication is written from and for a dominantly water-oriented perspective, it also inherently touches upon cross-cutting challenges that interact with the agricultural sector. We apply both perspectives in our analysis of the Communication.

2.3.1.1 History, Aims and Objectives

The Water Scarcity and Drought Communication represents and captures the milestones of EU policy to address water scarcity and drought through the Communication Document to European Parliament and the Council, titled "Addressing the challenge of water scarcity and droughts in the European Union" (European Commission 2007b; Kampragou et al. 2011). The Communication was adopted in 2008, after review. In the document, the Commission identified policy areas to address movement towards a water-efficient economy.

The 2007 Communication offers a variety of technical and political initiatives to mitigate the impacts of water scarcity and drought (Estrela and Vargas 2012). As part of the policy, the 2007 Communication put forth an initial set of policy options to address the challenges related to water scarcity and drought, with special emphasis on pricing, land use planning and water-saving (Kampragou et al. 2011).

The 2007 Communication also sets out a number of policy options addressing to increased drought frequencies as a result of climate change (Quevauviller 2014). The Communication also indicated the need for prioritization of drought risk management plans, the expansion of the European Drought Observatory, and a more rigorous use of the EU Solidarity Fund (Kampragou et al. 2011). The 2007 Communication on Water Scarcity and Droughts also address water pricing policies, advocating for 'putting the right price tag on water', 'allocating water more efficiently' and 'fostering water efficient technologies and practices'. These efficiency measures fit into the overall resource efficiency objective of Europe 2020.

2.3.1.2 Structure, Components and Implementation

A key feature of the water scarcity and drought policy is its use of a common framework. As a result, it lacks adequate differentiation of policy options and coping mechanisms for long-term or permanent discrepancies between water supply and demand (water scarcity) and temporary but sustained decrease in water availability as a result of natural forces (Kampragou et al. 2011).

Member states were encouraged to develop and implement Drought Management Plans (DMPs) as part of the Communication, considered to be an annex to the RBMP according to Article 13.5 of the WFD (Rossi 2009). RBMPs have to include a summary of the programmes of measures in order to achieve the environmental objectives (article 4 of WFD) and may be supplemented by the production of more detailed programmes and management plans (e.g. DMPs) for issues dealing with particular aspects of water management. The DMPs extend the criteria set forth by the WFD and aim to minimize impacts on the economy, social life and the environment, before the onset of drought using a multi-level approach (Rossi 2009).

Follow-up reports to the original Communication, which recalibrated priorities in 2008 and in 2010, revealed strides in water management efficiency. The follow-up reports also noted the limited response of member states in engaging in drought risk assessment, management, and development of DMPs (Kampragou et al. 2011).

The review of the Strategy for water scarcity and drought was integrated into the 'Blueprint to Safeguard European waters', in parallel with an analysis of the Implementation of the Water Framework Directive.

2.3.1.3 Relevance to Drought Policy Implementation

The 2007 Communication calls for a paradigm change from crisis-oriented to a planned drought risk management approach and expresses the need to explore all possibilities to improve water efficiency before exploring increase in supply (Estrela and Vargas 2012). The Communication also highlights the untapped potential for water efficiency measures in water user sectors, including agriculture, industry, distribution networks, buildings and energy production.

The Communication also states that clear, water use hierarchy established through participative approaches should inform policy-making (Estrela and Vargas 2012). More specifically, the Communication offers voluntary measures to cope with water scarcity and droughts, recommends development of DMPs and the establishment of a comprehensive European drought strategy, and discusses consideration of a European drought observatory (Estrela and Vargas 2012).

2.3.2 EC Communication 'Blueprint to Safeguard Europe's Water Resources'

The European Commission's Communication on 'Blueprint to Safeguard Europe's Water Resources' (European Commission 2012a) takes an inherently water-oriented perspective to its analysis. However, agriculture is also a major component of the recommendations set forth by the blueprint. The Communication underscores the increasing interplay between the water and agricultural sectors to address issues of water scarcity and drought.

2.3.2.1 History, Aims and Objectives

The 'Blueprint to Safeguard Europea's Water Resources' (European Commission 2012a) is an EU policy response to recent water challenges, to be encompassed by the EU 2020 Strategy and the Resources Efficiency Roadmap (Estrela and Vargas 2012). The Blueprint emerged from gaps in the WFD to address land use and management that affect both water quality and quantity. It assesses existing policy to highlight the obstacles and challenges which prevent actionable safeguarding of Europe's water resources. According to the Blueprint the main negative impacts on water resources are stemming from climate change, land use, economic activities, urban development and demographic change and are interlinked with each other.

As part of a larger goal to secure equal access to good-quality water, the primary objective of the Blueprint is to promote sustainable activities relating to water. The Blueprint does not aim to provide a one-size-fits-all solution and instead documents and assesses the vulnerability of EU waters. Within the Communication, it is argued that the objectives of the Blueprint are scattered throughout and already enshrined within the WFD, and consequently, the Blueprint attempts to gather, distill, and link the disparate elements of water policy as well as the root causes of negative impacts on water status.

The Blueprint identifies green growth as primary driving force behind changes to the water sector, with a special emphasis on innovation for water efficiency.

2.3.2.2 Structure, Components and Implementation

The Blueprint was developed in close cooperation with stakeholders, and is based on extensive public consultations. It consists of an overall Fitness Check of European Waters, as well as an assessment of the policies and measures in place.

2.3.2.3 Relevance to Drought Policy Implementation

The Blueprint identifies several pressures to address moving forward. First, the Blueprint calls for better implementation and increased integration of water policy objectives in policy areas such as the CAP, the Cohesion and Structural Funds, as well as energy, transport, and integrated disaster management. The Blueprint views the development of CIS guidance on natural water retention measures to be essential to facilitating such an integrated approach.

Second, the Blueprint identifies over abstraction of water. The Blueprint addresses the need for more quantitative water management, including identification and implementation of the concept of ecological flow, as well as a legal framework for addressing illegal abstraction of water.

As a reaction to these pressures, the Blueprint calls for the following measures:

- Addressing the vulnerability of EU waters: data from the Blueprint impact assessment show increasing trends in river flow droughts and flood-related losses in Europe over the last decades. The Blueprint encourages looking into measures based on an integrated disaster management approach, with special emphasis on extreme events including droughts.
- Increased financing measures under the CAP for (more) green infrastructure, specifically natural water retention measures.
- Continued development of the European Drought Observatory, an early-system aimed to serve as a preparedness measure for member states and affected stakeholders.
- Focusing on cross-cutting solutions, such EU water policy relating to Innovation Partnerships on Water and on Agricultural Productivity and Sustainability.

2.3.3 EU Water Framework Directive

The EU Water Framework Directive foregrounds much of the water policy field in Europe. As such, this section will pursue a strong focus on the water management perspective in which to ground discussions regarding implementation of water scarcity and drought policies within the WFD.

2.3.3.1 History, Aims and Objectives

The EU Water Framework Directive (WFD) (2000/60/EC) was adopted on 23 October 2000. It is the holistic legislation that encompasses all EU water policy, based on four main pillars that aim to first achieve 'good status' of all EU waters by 2015, second to establish river basin management plans, third to build a framework for integrated water management, and fourth, encourage public and stakeholder participation. The WFD operates on six-year cycles, with a new cycle set for 2015–2021.

Since 2000, the WFD has incorporated previous policies to create a single comprehensive framework for addressing surface waters, coastal waters, transitional waters, and groundwater, as well as linking to daughter directives that include: the Nitrates Directive (91/676/EEC), the Habitats Directive (92/43/EEC), the Urban Waste Water Treatment Directive (91/271/EEC), Integrated Pollution Prevention and Control Directive (96/61/EC), and other measures.

The WFD includes mandatory components for member states to implement. These instruments span several objectives, including costing/pricing, zoning of designated areas, abstraction and discharge permitting, and authorization of water quality-impacting activities (Kallis et al. 2005). The WFD is the dominant legislative instrument for addressing water-related issues.

2.3.3.2 Structure, Components and Implementation

According to the WFD, EU member states are required to develop a robust but flexible integrated water resources management system, based on the subsidiarity principles of river basin management planning (Quevauviller 2014). The provisions of the WFD imply that drought planning and management should be implemented at the level of river basins (Kampragou et al. 2011). Within the WFD, the River Basin Management Plans (RBMPs) act to prevent a drought crisis situation by clearly outlining the measure and actions to apply at varying triggering thresholds for water reserves (Kampragou et al. 2011). Therefore, drought scenarios must be clearly defined in the RBMP (Estrela and Vargas 2012). DMPs are contingency management plans supplementary to the River Basin Management Plans (RBMP)s. DMPs mainly aim to identify and schedule onset activation tactical measures to delay and mitigate the impacts of drought.

To implement the WFD, the Common Implementation Strategy (CIS) sets standards and guidance for implementation for all EU countries. Overall, the implementation strategy of the WFD is rather flexible and cooperative due to the vague core requirements set forth by the legal text of the WFD (Kallis et al. 2005). The WFD does not supply any direct financial support. However, funding opportunities for measures are available through EU Regional Policy and EU Common Agricultural Policy (CAP). The use of structural and cohesion funds, as part of the regional policy of the European Union, need to be mobilized by either municipality initiatives or water authorities of the MSs. Against the background of the financial crisis the EU took the temporary decision to improve the EU co-financing rates for selected MSs (Stanley et al. 2012).

Because the WFD is a more general framework, it was also implemented via other directives, including the Groundwater Directive and Nitrates Directive. It is also complemented by the Floods Directive and the Communication on water scarcity and drought.

The implementation of pricing instruments under the WFD is a way to provide an incentive for efficient water use. Water pricing serves not only as a powerful awareness-raising tool but also combines environmental with economic benefits.

(cp. European Commission 2012a). However, effective metering is a prerequisite for actualizing such incentive-based pricing instruments. Consequently, the WFD also makes use of mentions cost recovery of water services, operating on the polluter pays principle.

2.3.3.3 Relevance to Drought Policy Implementation

Though the WFD does not provide a common definition of drought (Estrela and Vargas 2012), nor does it explicitly address droughts (Quevauviller 2011), due to its innate flexibility as framework, the WFD does offer a dynamic, evolving strategy to address drought and water scarcity challenges in the context of climate change through planning processes (Kampragou et al. 2011).

In the context of drought, the WFD aims to provide technical tools and targeted guidance to member states on the best methods for incorporating and addressing drought risks in current and future management plans (Kampragou et al. 2011).

In addition, the WFD also provides general criteria for assessing the status of water bodies from a drought perspective (Estrela and Vargas 2012). Specifically, abstraction and discharge permitting is required by all member states (Landgrebe et al. 2011). Besides permitting, water efficiency targets, as outlined within the CIS based on water stress indicators, are also being implemented at the river basin level.

From the land management perspective, the WFD has also set up several mechanisms that work to prevent land degradation and desertification impacts, mainly through measures outlined in the Programme of Measures and the River Basin Management Plans provided by each member state. On the one hand, this allows for ample flexibility for adapting the measures to the ecological needs and boundaries of the local and regional ecosystems. However, this approach also often leads to differences in interpretation and implementation, and creates uneven levels of achievement with regards to drought measures.

In conclusion, though the WFD provides a flexible entry point for EU-wide operationalization of drought-related instruments and measures, there are still opportunities for improving its approach. As already touched on, the focus of the WFD on water quality and not water quantity leaves provisioning of the amount of water resources too general and insufficient to tackling issues of drought and water scarcity management. Moreover, the WFD and its daughter Directives (discussed in further sections), place a stronger emphasis on northern Europe, where there is limited need for measures relating to water quantity, so far. However, this is likely to change in the coming decades, with climate change impacts shifting water quantity regimes throughout the whole of Europe (CITE). In light of this, more expansive guidance for implementing specifically tailored measures for water scarcity and drought for regions within both southern and northern Europe is essential.

2.3.4 EU Floods Directive

The EU Floods Directive, as first glance, may not appear immediately relevant to understanding drought and water scarcity policy in Europe. However, the Directive is worth exploring from the water management perspective as a way to inform WS&D policy measures and policy strategies. There is a great deal of potential for harnessing overlaps between drought and flood policy instruments, which to date are minimal. The water, in particular the water efficiency, perspective thus offers a rich entry point for understanding interactions between the Floods Directive and WS&D.

2.3.4.1 History, Aims and Objectives

In 2007, following the increase in occurrence of floods throughout Europe, the Floods Directive (2007/60/EC) emerged as the principle body of policy for targeting flood risk management. By 2011, member states were asked to undertake a Preliminary Flood Risk Assessment to identify areas with significant flood risks. By 2013, member states were asked to prepare flood hazard and risk maps. As of 2015, member states were requested to prepare flood risk management plans with set objectives and methods for achieving those objectives.

2.3.4.2 Structure, Components and Implementation

The Floods Directive promotes an integrated and sustainable approach to the management of flood risk, in particular regarding the use of river basin-scale approaches that promote better environmental options of land use. By improving nature's water storage capacity and conserving water in natural systems, the severe effect of droughts and preventing floods are curbed (European Commission—DG ENV 2011). In order to harness synergies with the WFD, the Floods Directive Flood Risk Management Plans (FRMPs) should be coordinated with the River Basin Management Plans (RBMPs).

The implementation of the Floods Directive is based on a six-year planning cycle. In the first stage of the Floods Directive (already completed), preliminary flood risk assessment and the identification of areas of potential significant flood risk were largely based on available information about past significant floods and on forecasts. In the second stage of the Floods Directive risk management process was the production of flood hazard maps and flood risk maps for the areas identified as areas of potential significant flood risks by the end of 2013. The European Commission is currently assessing the information reported by member states (European Commission 2015).

The European Commission has performed a preliminary assessment of the implementation of the Floods Directive, which notes the diversity of approaches and methodologies used by member states.

2.3.4.3   Relevance to Drought Policy Implementation

Flood risk management plans may include the promotion of water efficiency practices, such as the improvement of water retention and controlled flooding. The Floods Directive is designed to improve green infrastructure and promote Natural Water Retention Measures (NWRM). Unfortunately, the direct impact of this directive to integrate and connect different planning purposes and scales, including drought is expected to be low. The reason for this is that most member states are prioritizing hard, technical flood protection measures for soft, non-technical ones. A lack of land availability is one out of many factors behind this development.

2.3.5   EU Habitats Directive and EU Birds Directive

Together, the EU Habitats and Birds Directives offer a strong conservation and land management perspective to inform water scarcity and drought challenges. Nature and agriculture are thus heavily referenced in analysis of its relevance to WS&D.

2.3.5.1   History, Aims and Objectives

Nature conservation and protection of biodiversity in the EU is regulated by two main directives: the Birds Directive (1979)[2] and the Habitats Directive (1992).[3] Both directives address the growing deterioration of natural habitats and increasing threats to wildlife species across the Europe. The overall objective of the two Directives is to ensure that the species and habitat types they protect are maintained, or restored, to a favourable conservation status (FSC) within the EU. The two directives do not directly reference WS&D and do not set explicitly WS&D-relevant obligatory requirements. However, the conservation measures outlined are inherently interlinked with issues of WS&D.

2.3.5.2   Structure, Components and Implementation

Member states are responsible for implementing the necessary laws, regulations, and administrative provisions to comply with both Directives, including: designation of Special Areas of Conservation (SACs) and establishment of necessary conservation measures for selected habitats, animal and plant species and designnatation of Special Protection Areas (SPAs) for targeted bird species with special

---

[2] Council Directive of 2 April 1979 on the conservation of wild birds (79/409/EEC).
[3] Council Directive 92/43/EEC of 21 May 1992 on the conservation of natural habitats and of wild fauna and flora.

conservation measures required in these areas. Natura 2000, the EU-wide ecological network of protected areas, encompasses the different types of conservation areas and serves as the cornerstone of the EU's action on nature conservation. Natura 2000 sites are therefore highly protected against damaging development.

For each Natura 2000 site, conservation measures, such as voluntary management plans (MPs) are either particularly designed for the site or integrated into other development plans. Alternative conservation measures include statutory, administrative or contractual measures and member states must choose at least one of the three categories (European Commission 2013), reported every six years (European Environmental Agency 2015).

In addition, the Habitats Directive asks member states to prepare Prioritised Action Frameworks (PAFs) (Art. 8) to set out the official nature conservation priorities for a country or region. The PAFs act as strategic planning tools encouraging access to as many EU financial instruments as possible in the financing of the Natura 2000 network (European Commission 2012b).

2.3.5.3 Relevance to Drought Policy Implementation

While damage to wildlife and habitats are few examples of direct impacts from drought and water scarcity (Wilhite et al. 2007; Vanneuville et al. 2012) the conservation measures for the protection of vulnerable species and habitats contribute to prevention and mitigation of the WS&D effects.

The designation of SACs and SPAs contributes indirectly to WS&D by way of necessary conservation measures. Additionally, the WFD ensures that protected areas (SACs and SPAs) of the Natura 2000 network are integrated into the river basin strategies. Such associated conservation measures might have a positive impact on the state of water systems, as they, for example, may prevent "the deterioration of natural habitats" (Art. 6.2 of the Habitats Directive) or by paying "particular attention to the protection of wetlands [...]" (Art. 4.2 of the Birds Directive). The Natura 2000 sites also work "to ensure a favourable conservation status of the habitat types and species, including all relations with their environment like water, air, soil and vegetation" (European Commission 2000; Sánchez Navarro et al. 2012 both in Vanneuville et al. 2012).

2.3.6 EU Groundwater Directive

To understand the EU Groundwater Directive and its interplay with WS&D, a water management perspective is important. As such, we focus on the water sector in exploring the Groundwater Directive in the context of drought.

2.3.6.1   History, Aims and Objectives

Within the larger European water policy framework, the Groundwater Directive (80/68/EEC), along with other similar Directives, is often referred to as daughter directives to the overarching WFD (Quevauviller 2014). Initially, the Groundwater Directive aimed to prevent the pollution of groundwater by hazardous substances and to check or eliminate the consequences of pollution already incurred.

The main goal of the Groundwater Directive is to ensure good water quality. Similar to the WFD, the Groundwater Directive focuses on water quality, rather than quantity.

2.3.6.2   Structure, Components and Implementation

The Groundwater Directive provides a binding agreement prohibiting any and all direct discharge of hazardous substances. Authorization, as well as a detailed record of the discharges has to be provided to the European Commission for both types of substances.

The new Directive, established in 2006, requires member states to establish quality standards to protect groundwater, based on identified appropriate levels based on local or regional conditions and thresholds. The standards are based on pollution trend studies, compliance, regular monitoring and reporting, and pollution reversal trends.

2.3.6.3   Relevance to Drought Policy Implementation

The Groundwater Directive is relevant to drought policy directly, as it aims to protect underground water reserves, by ensuring good water quality standards are upheld across all groundwater resources.

The Groundwater Directive focuses on water quality rather than quantity, which is still relevant to groundwater policy. As the need for groundwater aquifer monitoring will increase in coming decades and it is expected that the focus will increase more on water quantity.

2.3.7   European Common Agricultural Policy

As the name already suggests, the European Common Agricultural Policy (CAP) focuses primarily on the agricultural perspective to water scarcity and drought, though it also has implications within the water management perspective as well.

2.3.7.1   History, Aims and Objectives

Introduced in the 1950s, the CAP originally aimed to ensure a stable supply of food through improvements to agricultural productivity.

Beginning in the early 1990s, greater emphasis was placed on the environmental dimension with the introduction of agri-environment schemes in 1992. The Agenda 2000 reform established two pillars within the CAP, with the first pillar providing agricultural market and income support and the second pillar integrating rural development policy more broadly.

Since the 2000s, the CAP has undergone major reforms, including the introduction of decoupled farm payments and compulsory cross-compliance, both introduced in 2003.

The last round of the CAP reform for the current 2014–2020 programming period increases the links between the two pillars and thus offers a more holistic and integrated approach to policy support.

2.3.7.2   Structure, Components and Implementation

The structure of the CAP operates in seven-year budget cycles. Member states are awarded a degree of autonomy in applying the CAP at the national and regional level. This autonomy allows flexibility and results in varied implementation structures for both first and second pillars, and in effect, diverse impacts on soil across the EU.

The most recent CAP introduces a 'greening payment', where 30 % of the available direct payments national envelope is linked to the provision of particular sustainable farming practices. This means that in addition to the cross-compliance mechanism, a significant share of direct payments will in future be linked to rewarding farmers for the provision of environmental public goods.

Furthermore, under the second pillar a focus on environmental issues is enhanced with the provisions to allocate at least 30 % of the rural development programmes' budget to agri-environmental measures, organic farming or projects associated with environmentally friendly investment or innovation measures. The agri-environmental measures will need to complement greening practices, in this way meeting higher environmental protection targets. Furthermore, more focus is given to mainstreaming climate change mitigation and adaption actions, for example, by developing greater resilience to disasters such as flooding, drought and fire (European Commission—DG AGRI and Rural Development 2013). However, the most recent reform did not address the water issues explicitly.

Each member state is required to prepare Rural Development Programmes that in addition to specific agricultural development policies also address the challenges posed by the environment and climate change. Agri-environment schemes are the key measures for the integration of environmental concerns into the CAP. The RDPs encouraging farmers to conserve and enhance environmental features by providing incentives for the provision of environmental services.[4] In addition, a number of other rural development measures contribute to environmental protection (including water issues) and climate adaptation and mitigation measures, in particular, measures such as the cultivation of legumes, reduced use of fertilizers and pesticides, or organic farming methods.[5] Though indirectly, these measures also contribute to reduction of drought and water scarcity, as the improved soil quality also improves natural water retention capacity of soil. However, how rural development measures are designed and implemented is ultimately decided by the member states.

2.3.7.3 Relevance to Drought Policy Implementation

Water scarcity and droughts can cause significant economic impacts, particularly on agricultural activity, as it is one of the largest water demanding sectors after industry and domestic use (Farmer et al. 2008). As a major water user, agriculture plays a large role in impacting water scarcity and drought on ecosystems 'through effects such as the drying of wetlands, concentration of pollutants affecting river biota, increasing risk of forest fires, etc'. (Farmer et al. 2008).

The CAP remains one of the key EU policies relating to drought and water scarcity due to its scope and EU-wide coverage. Despite the reciprocal interlinkages between agricultural sector and water scarcity and droughts, the CAP only minimally offers financial and legal instruments to address drought (Rossi 2009).

However, the CAP remains the primary instrument for financial support for agriculture, and has in the past often led to increased pressures on water usage from this sector thus exacerbating the issue of water scarcity, in particular through the subsidies to water-intensive crops.

In addition, the definition of GAEC requirements at national level enables member states to address drought and water scarcity flexibly according to national priorities and local needs. One of three water-related GAEC standards focuses on water irrigation issues setting the requirement of compliance with authorisation procedures (European Community 2013) and is of relevance addressing water scarcity issues. The GAEC standards related to soil and carbon stock might be relevant to drought issues as well, as they support sustainable soil management practices that increase the resilience of farming systems to floods and droughts and contribute to soil health and quality in general. Such measures also improve natural water retention capacity of the soil and increase the resilience against drought.

The new CAP also offers a risk management toolkit as part of the rural development policy. The toolkit addresses adverse climatic events, through an income

---

[4]Agri-environment Measures, http://ec.europa.eu/agriculture/envir/measures/index_en.htm.
[5]Read more: http://www.ecologic.eu/9955.

stabilization tool that assists farmers with compensations paid for losses suffered as a result of adverse climatic events, such as severe drought. The water-related rural development measures focus mainly on water use and water pollution prevention and reduction measures. Therefore, it does not address the issue of drought prevention.

To address the issue of drought, both water and land ecosystems should be involved. Thus sustainable land management practices that increase the resilience of the farming systems have a large potential in contributing to drought prevention and reduction. In order to strengthen the aspect of risk prevention management of drought, coordination of activities between drought and agriculture policies should in addition to supporting improved water demand management practices, place a stronger focus on sustainable farming practices with potential to improve natural water retention capacity of soil. Several measures, including for example buffer strips, crop rotation, meadows and pastures, traditional practices, grasslands, terracing or green cover, can act as NWRM, by encouraging the retention of water within a catchment and, through that, enhancing the natural functioning of the catchment.[6]

2.4 Where to Go: A Conclusion on the Development of the European Perspective on Drought

Over the last two decades EU Drought Policy has developed from a series of scattered policies that range from broader forms of water governance that tackle water issues to more recently, direct policy actions to adapt and mitigate drought occurrences. In either case, the effectiveness of drought-related policy frameworks largely depend on the mobilization and operationalization of the policy through national and regional drought policies and initiatives (Bressers et al. 2013).

Moving forward, there is a critical need to shift from a crisis-oriented management approach to a risk-based (or even resilience-based) management approach. In addition, further integration and strengthening of various policy instruments, as suggested by the Blueprint for Safeguarding, that aim to promote policy measures, such as water efficiency, across water, land, and nature, and other management and policy spaces, are necessary to begin catalyzing such a shift. More support, in the form of financial mechanisms, at all policy levels is also essential, particularly within more complex policy mixes such as the WFD and the CAP. At the moment, the only policy instrument directly tackling drought that exists at the European level is the Europeans Commission's Communication on water scarcity and drought. However, due to its lesser status in relation to Directives, the Communication is still weak and lacks teeth in the policy landscape. Somehow coupling WS&D with the

---

[6]Natural Water Retention Measures, http://www.nwrm.eu/concept/3857.

Table 2.2 Policy instruments and strategies and their potential to contribute to European drought and water scarcity policies in the different environmental domains

Policies	Water			Nature	Agriculture
	Water supply	Water saving	Water allocation		
Water Framework Directive	◉	○	◉	○	○
Floods Directive	○	○	◉	○	○
Habitats Directive/ Birds Directive	○	○	◉	●	○
Groundwater Directive	●	○	○	○	●
Water Scarcity and Drought policy	◉	●	●	◉	●
Blueprint to Safeguard Europe's Water Resources	●	●	◉	◉	●
CAP	◉	●	◉	◉	●
Relevance for EU drought policy: ● high, ◉ medium, ○ low					

Floods Directive may offer one solution for upgrading the Communication while also supporting better integration of climate change-related events (please see Table 2.2).

In light of this, based on recommendations already set forth by Kampragou et al. (2011), we highlight several challenges that should be explored at the EU level. They include: (1) shifting from crisis management to risk management, (2) launching policy instruments and initiatives that promote water efficiency, (3) integrating environmental considerations when selecting drought mitigation actions, (4) increasing the knowledge base that informs policy instruments, (5) developing more holistic response and recovery frameworks and (6) harmonizing and disseminating policy instruments relating to drought and water scarcity.

Not surprisingly, an appropriate policy mix, consisting of a combination of mutually strengthening policy instruments, measures, and strategies is determining the success of European drought and water scarcity policies. This mix is ensuring the flexibility of policy responses that is needed to appropriately react to the water-related deterioration of land and water ecosystem caused by climate change.

References

Bressers H, O'Toole LJ Jr (2005) Instrument selection and implementation in a networked context. In: Elidas P, Hill MM, Howlett M (eds) Designing government. From instruments to governance. McGill-Queen's University, Montreal, pp 132–153

Bressers H, de Boer C, Lordkipanidze M, Özerol G, Vinke-De Kruijf J, Furusho C, La Jeunesse I, Larrue C, Ramos M-H, Eleftheria Kampa, Stein U, Tröltzsch J, Vidaurre R, Browne A (2013) Water governance assessment tool. With an elaboration for drought resilience. Report to the DROP project. CSTM University of Twente, Enschede

Council of the European Communities (1979) Council Directive of 2 April 1979 on the conservation of wild birds: (79/409/EEC). http://eur-lex.europa.eu/legal-content/EN/TXT/PDF/?uri=CELEX:31979L0409&from=EN. Accessed 17 Dec 2015

EC (2012) Communication from the Commission to the European Parliament the Council the European Economic and Social Committee and the Committee of the Regions A Blueprint to Safeguard Europe's Water Resources. COM (2012) 673 final, European Commission, Brussels, p 24

Estrela T, Vargas E (2012) Drought management plans in the European Union. The case of Spain. Water Resour Manag 26(6):1537–1553

European Commission (2000) Managing Natura 2000 sites: the provisions of Article 6 of the 'Habitats' Directive 92/43/CEE. http://ec.europa.eu/environment/nature/natura2000/management/docs/art6/provision_of_art6_en.pdf. Accessed 17 Dec 2015

European Commission (2007a) Accompanying document to the communication from the commission to the Council and the European Parliament on Addressing the challenge of water scarcity and droughts in the European Union—Impact Assessment. COM (2007) 414 final, Brussels

European Commission (2007b) Communication from the Commission to the European Parliament and the Council on addressing the challenge of water scarcity and droughts in the European Union. COM (2007) 414 final, Brussels

European Commission (2010) Second follow-up report to the communication on water scarcity and droughts in the European Union: COM (2010) 228 final. http://ec.europa.eu/environment/water/quantity/eu_action.htm. Accessed 14 Dec 2015

European Commission (2011) Report from the commission to the European Parliament and the council: third follow up report to the communication on water scarcity and droughts in the European Union COM (2007) 414 final. http://eur-lex.europa.eu/legal-content/EN/TXT/PDF/?uri=CELEX:52011DC0133&from=EN. Accessed 17 Dec 2015

European Commission (2012a) A blueprint to safeguard Europe's water resources. COM (2012) 673 final, Brussels http://eur-lex.europa.eu/legal-content/EN/TXT/?uri=CELEX:52012DC0673. Accessed 17 Dec 2015

European Commission (2012b) Format for a prioritised action framework (PAF) for Natura 2000: for the EU multiannual financing period 2014–2020. http://ec.europa.eu/environment/nature/natura2000/financing/docs/PAF.pdf. Accessed 17 Dec 2015

European Commission (2012c) Report on the review of the European water scarcity and droughts policy. Communication from the commission to the European Parliament, the Council, the European Economic and Social Committee and the Committee of the Regions: COM (2012) 672 final. http://www.semide.net/media_server/files/semide/topics/waterscarcity/report-review-european-water-scarcity-and-droughts-policy/Water_Scarcity_COM-2012-672final-EN.pdf. Accessed 14 Dec 2015

European Commission (2013) Agricultural policy perspectives brief N°5: Overview of CAP reform 2014–2020, Dec 2013, Brussels

European Commission (2015) Communication from the EC to the EP and Council, The WFD and the floods directive: action towards the good status of EU's water and to reduce flood risks. COM (2015) 120 final. http://eur-lex.europa.eu/legal-content/EN/TXT/PDF/?uri=CELEX:52015DC0120&from=EN. Accessed 16 Dec 2015

European Commission—DG AGRI and Rural Development (2013) Overview of CAP reform 2014–2020.   http://ec.europa.eu/agriculture/policy-perspectives/policy-briefs/05_en.pdf. Accessed 15 Dec 2015

European Commission—DG ENV (2007) Water scarcity and droughts in-depth assessment: second interim report. http://ec.europa.eu/environment/water/quantity/pdf/comm_droughts/2nd_int_report.pdf. Accessed 17 Dec 2015

European Commission—DG ENV (2011) Towards better environmental options for flood risk management.   http://ec.europa.eu/environment/water/flood_risk/pdf/Note_Better_environmental_options.pdf. Accessed 17 Dec 2015

European Commission—DG ENV (2012) Water scarcity & droughts: 2012 policy review—building blocks non-paper. https://www.bdew.de/internet.nsf/id/DE_Non-paper_on_Water_Scarcity_Droughts_-_2012_Policy_review_-_Building_blocks_April_2010/$file/Br%C3%BCssel_20100421_final.pdf. Accessed 17 Dec 2015

European Commission—JRC (2015) Reports of severe drought events: drought news in Europe. http://edo.jrc.ec.europa.eu/edov2/php/index.php?id=1051. Accessed 10 Dec 2015

European Community (2000) Directive 2000/60/EC of the European Parliament and of the Council of 23 October 2000 establishing a framework for Community action in the field of water policy: 2000/60/EC. http://eur-lex.europa.eu/legal-content/en/TXT/?uri=CELEX:32000L0060. Accessed 17 Dec 2015

European Community (2013) Annex II in Regulation (EU) No 1306/2013 of the European Parliament and of the Council of 17 December 2013 on the financing, management and monitoring of the common agricultural policy and repealing Council Regulations (EEC) No 352/78, (EC) No 165/94, (EC) No 2799/98, (EC) No 814/2000, (EC) No 1290/2005 and (EC) No 485/2008

European Environmental Agency (2009). Water resources across Europe—confronting water scarcity and drought. http://www.eea.europa.eu/publications/water-resources-across-europe/at_download/file. Accessed 1 Dec 2015

European Environmental Agency (2010) Water resources: quantity and flows—SOER 2010 thematic assessment. http://www.eea.europa.eu/soer/europe/water-resources-quantity-and-flows. Accessed 17 Dec 2015

European Environmental Agency (2015) State of nature in the EU: results from reporting under the nature directives 2007–2012. http://www.eea.europa.eu/publications/state-of-nature-in-the-eu/download. Accessed 10 Dec 2015

European Union (2010) Water scarcity and drought in the European Union. http://ec.europa.eu/environment/water/quantity/pdf/brochure.pdf. Accessed 17 Dec 2015

Farmer A, Bassi S, Fergusson M (2008) Water scarcity and droughts: European Parliament. Policy Department Economic and Scientific Policy. http://www.europarl.gr/resource/static/files/projets_pdf/water_scarcity.pdf. Accessed 12 Dec 2015

Flanagan K, Uyarra E, Laranja M (2011) Reconceptualising the 'policy mix' for innovation. Res Policy 40(5):702–713

Flörke M, Wimmer F, Laaser C, Vidaurre R, Stein U, Tröltzsch J, Dworak T, Stein U, Marinova NJF, Ludwig FSR, Giupponi C, Bosello F, Mysiak J (2011) Climate adaptation: modelling water scenarios and sectoral impacts. Final Report. http://climwatadapt.eu/sites/default/files/ClimWatAdapt_final_report.pdf. Accessed 28 Oct 2011

Informal Council of Environment Ministers (2007) Water scarcity and drought (WS&D). https://www.google.de/url?sa=t&rct=j&q=&esrc=s&source=web&cd=1&cad=rja&uact=8&ved=0ahUKEwjk6aG6xLrJA-hVFeQ8KHevJBrYQFggfMAA&url=http%3A%2F%2Fwww.semide.net%2Ftopics%2FWaterScarcity%2Fbackground%2FWSDBackgroundext25jul07_EN.pdf%2Fdownload%2F1%2FWSDBackgroundext25jul07_EN.pdf&usg=AFQjCNEAve740bAHcROftzwHifgQNgG9Aw&sig2=2HwoDRjXn-Lq2n_ilF4Wi-g&bvm=bv.108194040,d.ZWU. Accessed 17 Dec 2015

Jol A, Šťastný P, Raes F, Lavalle C, Menne B, Wolf T (2008) Impacts of Europe's changing climate: 2008 indicator-based assessment. European Environment Agency. http://www.eea.europa.eu/publications/eea_report_2008_4. Accessed 17 Dec 2015

Kallis G, Briassoulis H, Liarikos C, Petkidi K (2005) Integration of EU water and development policies: a plausible expectation? In: Briassoulis H (ed) Policy integration for complex environmental problems: the example of mediterranean desertification. Ashgate, Burlington, Vt

Kampragou E, Apostolaki S, Manoli E, Froebrich J, Assimacopoulos D (2011) Towards the harmonization of water-related policies for managing drought risks across the EU. Environ Sci Policy 14(7):815–824

Kaufmann-Hayoz R, Bättig C, Bruppacher S, Defila R, Di Giulio A, Flury-Kleubler P, Friederich U, Garbely M, Gutscher H, Jäggi C, Jegen M, Mosler H-J, Müller A, North N, Ulli-Beer S, Wichtermann J (2001) A typology of tools for building sustainability strategies. In: Kaufmann-Hayoz R, Gutscher H (eds) Changing things—moving people. Birkhäuser Basel, Basel, pp 33–107

Kazmierczyk P, Reichel A, Bakas I, Isoard S, Kristensen P, Werner B, Jol A, Kurnik B, Fogh Mortensen L, Ribeiro T, Adams M, Lükewille A, Christiansen T, Meiner A, Jones A, van Aardenne J, Fernandez R, Spyropoulou R, Collins R (2010) The European environment—state and outlook 2010: water resources—quantity and flows. http://www.eea.europa.eu/soer/europe/water-resources-quantity-and-flows. Accessed 17 Dec 2015

Landgrebe R, Naumann S, Cavalieri S, Wunder S (2011) Policy context and policy recommendations for LEAD in cropland—general: Deliverable 141, http://www.envistaweb.com/leddris/. Accessed 16 Dec 2015

Mysiak J, Maziotis A (2012) Economic policy instruments for sustainable water. Presentation at the 1st Pan-EU drought dialogue forum Cyprus. http://www.eu-drought.org/media/default.aspx/emma/org/10815374/7.+EPIWATER_Maziotis__31Oct2012.pdf. Accessed 1 Dec 2015

Portuguese Ministry of Environment, Spatial Planning and Regional Development (2007) Water scarcity and drought. A priority of the portuguese presidency. ftp.infoeuropa.eurocid.pt/files/database/000042001-000043000/000042875.pdf. Accessed 12 Dec 2015

Quevauviller P (2011) Adapting to climate change: reducing water-related risks in Europe—5EU6 policy and research considerations. Environ Sci Policy 14(7):722–729

Quevauviller P (2014) European water policy and research on water-related topics—an overview. J Hydrol Part B 518:180–185

Reichardt K, Rogge K (2015) How the policy mix and its consistency impact innovation: findings from company case studies on offshore wind in Germany. Working Paper Sustainability and Innovation (7):1–37

Rossi G (2009) European Union policy for improving drought preparedness and mitigation. Water Int 34(4):441–450

Sánchez Navarro R, Schmidt G, Benítez Sanz C, Dubar M, Hernández T, José M, Raso Quinta J, Seiz Puyuelo R (2012) Environmental flows as a tool to achieve the WFD objectives:

Discussion Paper. http://www.pianc.org/downloads/euwfd/Eflows%20discussion%20paper%20draft%20NAVI%20TG%20response%20v2%2010-8-12.pdf. Accessed 15 Dec 2015

Schmidt G, Benítez JJ, Benítez C (2012) Working definitions of water scarcity and drought. https://circabc.europa.eu/sd/d/02a234f7-ac60-4f81-bd8d-a3a0973e77d1/55171-Drought-WS_Definitions_V4-27Abril2012.doc. Accessed 12 Dec 2015

Stanley K, Depaoli G, Strosser P (2012) Comparative study of pressures and measures in the major river basin management plans in the EU. Task 4b. Financing water management and the economic crisis—a review of available evidence. ACTeon, France

Tallaksen LM (2007) Tørke–også i ikke-tørrestrøk? Presentation at Faglig-Pedagogisk Dag (2007) www.geo.uio.no/for_skolen/ (cited not accessed)

Vanneuville W, Werner B, Uhel R (2012) Water resources in Europe in the context of vulnerability: EEA 2012 state of water assessment. EEA report 11/2012. Office for Official Publications of the European Union, Luxembourg

Vogt J (2011) Developing the European Drought Observatory (EDO). http://www.isprambiente.gov.it/files/ec-expert/vogt.pdf

Vogt J, Barbosa Ferreira P, Hofer B, Singleton A (2011) Developing a European drought observatory for monitoring, assessing and forecasting droughts across the European Continent. In Geophysical Research Abstracts (13)

Water Scarcity and Droughts Expert Network (2007) Drought management plan report: including agricultural, drought indicators and climate change aspects. http://www.magrama.gob.es/imagenes/en/090471228012641f_tcm11-17923.pdf. Accessed 12 Dec 2015

Wilhite DA, Svoboda MD, Hayes MJ (2007) Understanding the complex impacts of drought. A key to enhancing drought mitigation and preparedness. Water Resour Manage 21:763–774

Wilhite DA, Sivakumar MV, Pulwarty R (2014) Managing drought risk in a changing climate: The role of national drought policy. Weather Clim Extremes 3:4–13

Chapter 3
The Governance Assessment Tool and Its Use

Hans Bressers, Nanny Bressers, Stefan Kuks and Corinne Larrue

3.1 Introduction: The Implementation Challenge

Especially in the context of climate change adaptation, the sustainability of natural resources requires an integrative vision, apt policies and adequate implementation to realise the proposed measures in practice. Mentioning all three issues does not imply that they have clear boundaries between them. Instead, between vision and policies and between policies and their implementation mutual influences occur. Often policies get further shape in the process of implementing them. This is more true when the policy formation and implementation have a multi-actor character, like in most cases of drought resilience management in Northwest Europe. Instead of singular policies with a separate implementation process, drought management is often a combination of water system and behavioural adaptations, which relates to and draws resources from various policy sectors, and requires concerted action on multiple levels and time scales (Bressers and Lulofs 2010). Such a challenge can be labelled as complex and dynamic. It requires a lot of 'connective capacity' (Edelenbos et al. 2013). It is essentially this nature of 'complex and dynamic multi-actor interaction processes' that requires a good governance context to enable the realisation of practice projects. Without a good governance context the degree of trust, openness and mutual liking is likely too low to allow for real cooperation.

In Chap. 1 it has been explained that the analyses in this book make use of a specific theory and method of governance analysis that aims to be practice oriented in

H. Bressers (✉) · S. Kuks
CSTM, University of Twente, PO Box 217, 7500 AE Enschede, The Netherlands
e-mail: Hans.bressers@utwente.ul

N. Bressers · S. Kuks
Water Authority of Vechtstromen, PO Box 5006, 7600 GA Almelo, The Netherlands

C. Larrue
Université Rabelais de Tours, 60 Rue du Plat d'Étain, 37000 Tours, France

© The Author(s) 2016
H. Bressers et al. (eds.), *Governance for Drought Resilience*,
DOI 10.1007/978-3-319-29671-5_3

that it tries to assess to what degree the governance context is supportive or restrictive for the realisation of the chosen policies and projects. In this chapter, this so-called Governance Assessment Tool will be presented and explained. Also some remarks guiding its use will be made. In the next section we will first introduce the theoretical approach in which the Governance Assessment Tool is rooted (Sect. 3.2). In this so-called Contextual Interaction Theory, operational decision making and implementation processes are studied from three actor characteristics (motivations, cognitions and resources) that are influenced by the three contextual layers of case specific circumstances, the governance context and the more general wider context like the technological development. Section 3.3 presents the Governance Assessment Tool itself, including its descriptive and evaluative questions. Thereafter in Sect. 3.4 we will guide the reader on how to use it to facilitate governance analysis.

3.2   Understanding Policy Implementation as Multi-actor Interaction Process: Contextual Interaction Theory

The Governance Assessment Tool is rooted in a theory of policy implementation that is labelled Contextual Interaction Theory (Bressers 2004, 2009; De Boer and Bressers 2011). It views implementation processes not top down, as just the application of policy decisions, but as multi-actor interaction processes that are ultimately driven by the actors involved. Thus it makes sense to explain the course and results of the process from that simple starting point and to place these actors and their main characteristics central stage in any analytical model. This is also relevant because in the history of implementation research hundreds of crucial success factors were proposed and used to analyse all kinds of different cases. This can be theoretically interesting when one can try to carve out the impact of a single factor from those of all the others. In practical reality however practitioners must deal with situations in which all factors are around simultaneously, and thus with combinations of all factors that are thought to matter (Bressers and O'Toole 2005). Even in a rather simple model of fifteen factors having each only two possible values there are some thirty thousand different combinations of circumstances that can be imagined. That is not only unworkable as an analytical tool (Goggin 1986), it is also overdone. There are no thirty thousand (or more) fundamentally different implementation settings. But since interaction processes are human activities, all influences flow via the key characteristics of the actors involved (Bressers and Klok 1988). Thus, it is possible to explain the course and effects of implementation processes with a set of three core factors per actor. Such explanatory model is far more parsimonious, at least to begin with. All other factors, including governance conditions, are regarded as belonging to the context that may influence this set of core factors. In Fig. 3.1 we include these factors: their motivations that may spur the actors into action, their cognitions, information held to be true, and their resources, providing them with capacity to act individually and power in relation to other actors. Among the actors involved in the process there need to be a sufficiently

Fig. 3.1 Process model with the actor characteristics used in Contextual Interaction Theory (*Source* Bressers 2009)

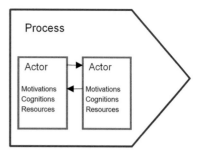

strong combination of motivations, cognitions and resources to enable the process to succeed (Bressers 2004; Owens 2008).

The basic assumptions of Contextual Interaction Theory are quite simple and straightforward.

The theory's main assumptions are:

1. Policy processes are multi-actor interaction processes. Both individuals, often representing organisations or groups, or organisations themselves can be considered actors when participating in the process.
2. Many factors may have an influence but only because and in as far as they change relevant characteristics of the involved actors.
3. These characteristics are: their motivation, their cognitions and their resources, providing them with capacity and power (Knoepfel et al. 2011: 68).
4. These three characteristics are influencing each other, but cannot be limited to two or one without losing much insight (Mohlakoana 2014).
5. The characteristics of the actors shape the process, but are in turn also influenced by the course and experiences in the process and can therefore change during the process. There is a dynamic interaction between the key actor characteristics that drive social interaction processes and in turn are reshaped by the process. Deliberate strategies of actors involved can try to promote such changes both in other actors and within their own group or organisation.

And, as we will discuss further on in this section:

6. The characteristics of the actors are also influenced by conditions and changes in the specific case context of for instance characteristics of the geographical place and previous decisions that among others can set the stage for some actors and exclude others from the process.
7. A next layer of context is the structural context of the governance regime. This is the context that our Governance Assessment Tool concentrates on.
8. Around this context there is yet another more encompassing circle of political system, socio-cultural, economical, technological, and problem contexts. Their influence on the actor characteristics may be both direct and indirect through the governance regime.

Figure 3.2 shows these various layers of context. They are pictured as over-lapping circles that all three have direct potential impact on the characteristics of the actors, indicating that wider context do not need to first influence governance and then the specific case context before having an impact, even though some of their influence will work like that. Also the other way around, the case process influencing the evolution of the contexts, is possible, but this influence will mostly be limited to the specific context.

In Fig. 3.3, many theorems and other ideas are employed. Compared to Fig. 3.1 this figure does also show process development (change processes—in the form of the processes over time). The actor characteristics are much more elaborated here, not visualised as linked to specific actors and for presentation reasons placed outside of the process box. This enables showing the mutual influences between these factors and the process itself (compare Mohlakoana 2014).

Motivations—The origins of motivations for behaviour, including for the positions taken in interaction processes, first of all lay in own goals and values. Self-interest, like in many economic theories, plays of course a strong role here. But also more altruistic values can directly lead to genuine own goals (Gatersleben and Vlek 1998). External pressures can be also a motivating force. Like all motivational factors they could in principle also be conceptualised as belonging to one's 'own' purposes. However, such conceptualisation will make them often forgotten or underemphasised. These external pressures can be based on force, but more often will be softer influences from normative acceptance of the legitimacy of such external wishes and even by identification with the group from which such expectations come. Last but not always least also the 'self-effectiveness assessment' (Bandura 1986) can play a large role as a motivational factor. This concept points to the de-motivational effect that occurs when an actor perceives its preferred behaviour as beyond its capacity. It shows part of the relation between motivation and the availability of resources. While all of these elements are rooted in social and

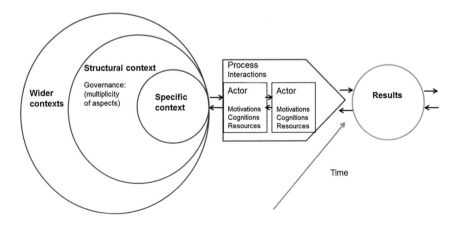

Fig. 3.2 Interaction processes influenced simultaneously by various layers of context

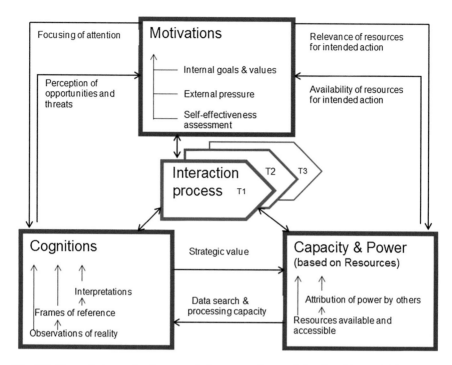

Fig. 3.3 Dynamic interaction between the key actor characteristics that drive social interaction processes and in turn are reshaped by the process (*Source* Bressers 2009)

learning psychology and thus seem to apply to individuals, also organisations and individuals representing groups and organisations (so-called 'corporate actors') can have the same set of origins of motivations.

Cognitions—The cognitions of actors (interpretations of reality held to be true) are not only a matter of observations and information processing capacity, though these aspects are important and with the information technology revolution are a source of quick changes. In policy sciences the so-called 'argumentative turn' (Fischer 1995; Fischer and Gottweis 2012), reflects a variety of approaches that emphasise that knowledge is produced itself in mutual interactions, based on interpretations of reality of actors, that themselves are mediated by frames of reference. Some frames of reference are termed by Axelrod (1976) as 'cognitive maps', by Schön (1983), Schön and Rein (1994) and later van Hulst and Yanow (2014) as 'frames', by Sabatier and Jenkins-Smith (1999) as 'policy core beliefs' and 'deep core beliefs'. Dryzek (1997) speaks of 'discourses', thereby also stressing the language dependency of understanding and the role of words, one-liners, stories and the like to guide, but also to restrict and bias understanding. While these approaches are quite different in their conceptual understanding and methodology of reconstruction, they also share some understandings: that cognitions are not just factual information about, but more interpretations of reality, and that such interpretations are influenced by filters, frames and interactions with other actors. Not the whole of the theoretical approaches mentioned, but only this 'common ground' is incorporated in the cognitions box of the contextual interaction theory. Part of these frames is related to 'boundary judgments', ideas about what does and what does not belong to a subject at hand (Bressers and Lulofs 2010). These different cognitions can refer to subjects or aspects of the project or the problems it wants to solve, or about the time frames that are relevant like short term results versus contributing to a long term vision, or about the relevant levels and scales: just local, or also embedded in a higher level or bigger scale of problem-solving. Differences in these boundary judgments between the various actors in the process can have significant impacts on their interactions in the process.

Resources—While resources as an actor characteristic are important to provide capacity to act, in the relational setting of an interaction process they are also relevant as a source of power. Resources are here meant to be any asset that public and private actors can use to support their actions. This implies that the relevance of resources is dependent on the actions an actor wants to perform. Having resources that other actors need access to for their preferred actions provides a basis of power. While in the previous figures the actor characteristics are purely linked to separate actors in Fig. 3.3 they are related to the actions and interactions in the process. Therefore this box is labelled "capacity and power" in Fig. 3.3. The relationship between power and resources is not always direct. Power in first instance largely results of attribution to an actor by others. However when this attribution is not backed by real resources it is fragile as soon as it is challenged. The resources that are the root of these powers encompass much more than formal rules, though legal rights and other institutional rules can be an important part of it, aside from resources such as money, skilled people, time and consensus (Klok 1995; Knoepfel et al. 2011).

Not only the resources of the actors themselves, but moreover the dependency of an actor on the resources of another actor shapes the balance of power. A classic example is the dependency of authorities on the jobs created by industry, which industry can use as a source of negotiation power. The example also shows that in Contextual Interaction Theory not just formal powers count, but that power can be based on all kinds of resources. Resources not only shape power relations, but are also a prerequisite for action as such, determining the capacity to act of any actor. The resource base for action can be greatly enlarged by engaging in dependencies with other actors with relevant resources, at the expense of loss of autonomy and thus—in some cases—power. Whether a specific resource contributes to capacity and power depends on the action that is intended. Resources that seem irrelevant to get certain things done might be essential to get other things done.

There are mutual relations between the three key actor characteristics. Every change in one of the three has influences on the other two. While we typically start with mentioning motivation, many would like to start with the way reality is understood and problems and chances perceived, or whether some useful infor-mation is available (on relevant technology, economics, social, geographical or environmental conditions), as a prerequisite for motivation. It must be borne in mind that the influence is mutual: without certain interests and values, available data may be overwhelming and too time consuming to process. The development of information needs some focusing of attention (creating selective perception as a bias). The actions for which an actor is motivated require resources, and the availability of those resources is bound to influence the actors' ambition, for instance because a lack of necessary resources creates a low self-effectiveness assessment (Bandura 1986). While 'knowledge is power' (attributed to Francis Bacon 1561–1626) may be one-sided, it is certainly true that information can serve strategic purposes and hence can be used as one of the bases of power. On the other hand the gathering and processing of data is also an activity that needs resources. Lastly, the three factors are not only shaping, but are also (re)shaped by the activities and interactions that happen in the process.

Above we explained the model of social interaction processes in Contextual Interaction Theory. It is applied to the implementation of drought resilience measures in this book. It has been used extensively in implementation case studies on various fields, also outside of the water sector. Its treatment in this section served to show what our understanding is of the nature of the processes that may find more or less supportive governance and other contexts in practice. Contextual Interaction Theory contains not only this part on the interaction process of implementation and realisation but also a part on these contexts and their relevance. In the next section we will explain the layers of context further and how they may be supportive or restrictive for the success of the interaction processes under study. In doing this we will concentrate on the layer of governance characteristics.

Governance is often said to differ from earlier developed concepts like government or policy in that it emphasises the multi-level and multi-actor character of all forms of steering of any specified (sub)sector of society. In our approach to the concept of governance we do not only discern the multiplicity of the levels and of the actors involved, but also apply the idea that the concept of governance assumes multiplicity to the dimensions of the older concept of policy: goals, instruments and the means to apply them (Howlett 2011). In each governance context there will likely be multiple goals involved, multiple instruments and multiple means to apply them. In Chap. 1 it was explained how this led to a conceptualization of governance in five dimensions (Bressers and Kuks 2003).

In Fig. 3.4 these dimensions are listed as filling the structural layer of context for the implementation processes. The structural context at for instance national level is much more stable than the specific case context. The structural context will to a far lesser degree be influenced back by individual implementation cases. In fact it is the essence of the difference between the specific and the structural context that the latter holds for in principle all similar cases and not only for any specific case. This is not to say that the structural context is not changing over time, just that these changes are even more the emergent result of many actors and factors than changes in the specific context.

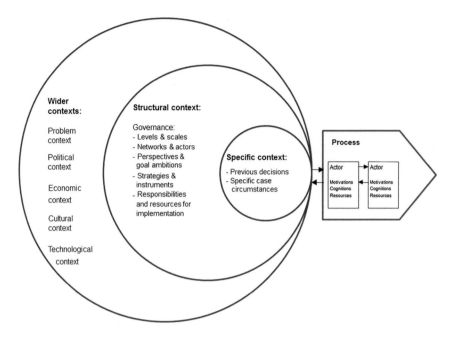

Fig. 3.4 Elaboration of the layers of context in contextual interaction theory (*Source* Bressers 2009)

3.3 The Governance Assessment Tool

The previous section has explained some theoretical roots of the type of governance assessment that we use in this book. In this section we will explain the Governance Assessment Tool that has been further developed in the context of the DROP project. To be able to systematically describe what the five dimensions of governance look like in the given governance context we developed a set of questions that can be used to guide the analysis of policy and other archival documents, and structure the conduct and analysis of qualitative interviews with key informants. Figure 3.5 gives an overview. All five dimensions include a descriptive question regarding the time dimension—that is, 'Have any of these changed over time or are likely to change in the foreseeable future'. In the context of the DROP project, it was particularly relevant to include this time dimension to spot any visible trends in the governance dimensions across case study regions. This is particularly important in Europe where countries face the same deadlines, like the 2015–2021–2027 assessment years of the Water Framework Directive.

While it is not difficult to see that all five elements of governance have strong relevance for the inputs into the process and the motivations, cognitions and resources of the actors therein, they do not specify what aspects of them create a more or a less stimulating context for the process.

Governance dimension	Main descriptive questions
Levels and scales	Which administrative levels are involved and how? Which hydrological scales are considered and in what way? To what extent do they depend on each other or are able to act productively on their own? Have any of these changed over time or are likely to change in the foreseeable future?
Actors and networks	Which actors are involved in the process? To what extent do they have network relationships also outside of the case under study? What are their roles? Which actors are only involved as affected by or beneficiaries of the measures taken? What are the conflicts between these stakeholders? What forms of dialogue between them? Are there actors with a mediating role? Have any of these changed over time or are likely to change in the foreseeable future?
Problem perspectives and goal ambitions	Which various angles does the debate of public and stakeholders take towards the problem at hand? What levels of possible disturbance are current policies designed to cope with? What levels of disturbance of normal water use are deemed acceptable by different stakeholders? What goals are stipulated in the relevant policy white papers and political statements? Have any of these changed over time or are likely to change in the foreseeable future?
Strategies and instruments	Which policy instruments and measures are used to modify the problem situation? To what extent do they reflect a certain strategy of influence (regulative, incentive, communicative, technical etc.)? Have any of these changed over time or are likely to change in the foreseeable future?
Responsibilities and resources	Which organisations have responsibility for what tasks under the relevant policies and customs? What legal authorities and other resources are given to them for this purpose or do they possess inherently? What transparencies are demanded and monitored regarding their use? Is there sufficient knowledge on the water system available? Have any of these changed over time or are likely to change in the foreseeable future?

Fig. 3.5 Main descriptive questions specifying the five elements of governance for water management implementation (*Source* Bressers et al. 2013)

To indicate what status of the five elements of governance contributes to a stimulating rather than restrictive governance context for the implementation and realisation of water management measures, four quality criteria have been elaborated over the years (Bressers and Kuks 2004; De Boer and Bressers 2011; Kuks et al. 2012; Bressers et al. 2013, 2015). The structural context influences the process not only through its direct contents, but also through its *extent* and *coherence* (Knoepfel et al. 2001, 2003; Bressers and Kuks 2004). The extent refers to the completeness of the regime. The coherence is the degree to which the various elements of the regime are strengthening rather than weakening each other.

Regimes with an insufficient *extent* are by definition weak as guardians of sustainable use of water resources, while some relevant parts of the domain go unregulated. Most of the time, growing complexity is an answer to real needs and developments. As a matter of fact, societies in modern times have generally grown into a situation of increased complexity. Increased populations, borders, overlaps, activities, rivalries, etc. are a fact of our current living environments. A growing complexity in governance can be viewed as a logical adaptation to that development (Gerrits 2008; Teisman et al. 2009). Many external change agents, such as technological developments, add new scales, new actors, new problem perceptions, new instruments, and new responsibilities to the existing ones. The essence of extent is not the number of involved scales, actors, perceptions, instruments and resources as such, but rather the degree to which these are complete in reflecting what is relevant for the policy or project. In that sense it should not be mistaken for another way of making a descriptive inventory like with the descriptive questions.

By *coherence* we mean the following: When more than one layer of government is dealing with the same natural resource (as is often the case), then coherence means inter alia that the activities of these layers of government are recognised as mutually dependent and influencing each other's' effects. Likewise if more than one scale is relevant the interaction effects between those scales should be considered. When more than one actor (stakeholder) is involved in the policy, coherence means that there is a substantial degree of interaction in the policy network, and preferably productive interaction providing coordination capacity. When more than one use or user is causing the problem of unsustainable resource use for example, coherence means that the various resulting objectives are analysed in one framework so that deliberate choices can be made if and when goals and/or uses are conflicting. When the actors involved have problem perceptions that start from different angles, coherence means that they are capable of integrating these to such an extent that a common ground for productive deliberation on ambitions is created. The same holds for instrumental strategies that are used to attain the different objectives, as well as for the different instruments in a mix to attain one of these objectives. Coherence of the organisation of implementation means that responsibilities and resources of various persons or organisations that are to contribute to the application of the policy are co-ordinated, or these actors themselves are co-ordinated.

In the implementation process, the additional fragmentation that is typical for complex but non coherent regimes will tend to lead to more discord between the actors (goals), more uncertainty (cognitions), and more stalemates (power) and, thereby, can hamper implementation. In the implementation process, coherence of the structural context (the regime) will tend to lead to less discord (due to more 'win-win'—solution creativity), less (subjective) uncertainty (due to more exchange of information and less distrust) and less stalemates (due to less possibilities for target groups to play the implementers off against each other and more standard operation procedures for the solution of conflict).

While in stable and relatively simple situations extent and coherence might be sufficient to evaluate the degree to which the governance context is supportive or restrictive for the implementation of policies and projects, more complex and dynamic situations require extra criteria (De Boer and Bressers 2011). For the success and failure of complex spatial projects and policy implementation in complex situations in general, some form of 'adaptive implementation' is important, trying not only to see the reality as a field of obstacles, but also as a terrain of potential—often unexpected—opportunities and being adaptive enough to use every 'window of opportunity' to bring the ultimate purpose closer to realisation. Therefore it is essential that the somewhat static factors of extent and coherence are supplemented with the factor of flexibility, indicating to what degree the relevant actors have formal and informal liberties and stimuli to act.

Flexibility is defined here as "the degree to which the regime elements support and facilitate adaptive actions and strategies in as far as the integrated (et al. multi-sectoral) ambitions are served by this adaptiveness". Consequently it is also the degree to which hindrances for such adaptive behaviour are avoided. Like extent and coherence, the flexibility of the regime as such could be understood in terms of the five elements of governance described above. A regime is more flexible in as far as the relationships between the levels and scales involved are more based on decentralisation of power, without upper levels withdrawing support. This is closely related to empowering rather than controlling relations, and thus on trust. A similar feature describes flexible regimes in terms of actor relations in the policy network. Here too the combination of giving leeway to each actor group to optimise its contribution to the whole programme while still viewing the programme as a joint effort qualifies as flexibility. In terms of general problem perception and goal ambitions flexibility implies that these in their variety are not only integrated into a sort of common denominator (like with coherence), but also that these mixtures are allowed to be different in emphasis according to the opportunities of the context in the various concrete situations. This implies some acceptance of uncertainty and openness to emergent options, which again relates to trust. The instruments and their combinations in policy strategies or mixes are more flexible in as far as means from different sources (like public policies and private property rights) may be used as well as indirect means (here relating to opening or improving options for the use of means that more directly serve the goals) are available and allowed to be used. Lastly the flexibility of the organisation responsible for the implementation—the responsibilities and resources given by the policy programme(s)—can be measured by the discretion available to pool resources like funds and people with those of others to serve integrated projects and to be held accountable on the basis of the balanced virtues of the achievements (as in an integrated project), rather than on the basis of separate performance criteria.

Given the dynamic and change oriented nature of some policies, like realising more drought resilience in the water system, there is yet another regime quality that can be influential for the practical process. That is the obvious, but no less important aspect of intensity. *Intensity* is "the degree to which the regime elements urge changes in the status quo or in current developments". The 'amount of change' is thereby measured in analogy with Newton's 'law of inertia', so as the degree of energy it takes to produce the change. In systems theory, induced changes will typically meet negative feedback loops, weakening their impact, while in some cases positive feedback loops creating dynamics for permanent change are also conceivable (True et al. 1999; Bressers and Lulofs 2009). In policy studies' terms intensity is related to the size of the task to create new dynamics by creative cooperation, or conflict. Consequently this urges change of conservative motivations or overcoming them by power, changing cognitions including widening of boundary judgments regarding the issues at stake, and developing new availabilities and combinations of resources. In other words: with more intensity the urge to use clever adaptive strategies to deal with and change the setting of the process increases. In terms of the five elements of governance intensity is greater in as far as also upper levels are more deeply involved, actors that are also powerful in other domains are more deeply involved in the relevant policy network for the issue at stake, the issue plays a larger role in the public debate leading to a greater openness to try to push developments away from a business-as-usual track (thus with more ambitious goals), the instruments made available to be used include more interventionist ones, and the amount of resources made available for implementation is larger.

In summary, the four criteria are defined by the questions that they pose:

1. Extent: are all elements in the five dimensions that are relevant for the sector or project that is focused on taken into account?
2. Coherence: are the elements in the dimensions of governance reinforcing rather than contradicting each other?
3. Flexibility: are multiple roads to the goals, depending on opportunities and threats as they arise, permitted and supported?
4. Intensity: how strongly do the elements in the dimensions of governance urge changes in the status quo or in current developments?

For each of the five dimensions of governance, the four criteria mentioned above are specified with specific questions (Fig. 3.6) which forms a matrix of assessment for the governance of drought and water scarcity for a region. This matrix forms the core of the Governance Assessment Tool (GAT). Together, these questions shed light on the degree of supportiveness or restrictiveness of the governance context towards the implementation of policies and projects. It is important to note that the GAT does not assess the functioning or success of an actor or a specific adaptation plan. Rather, the GAT assesses the entire governance context, enabling reflections on the way that this context supports or restricts the implementation of policies and projects.

While this version is developed by the scientists of the "governance team" in the DROP project its usability reaches far beyond drought management. In fact the tool is applicable to a wide range of implementation projects in water management and even beyond.

Governance dimension	Quality of the governance regime			
	Extent	Coherence	Flexibility	Intensity
Levels and scales	How many levels are involved and dealing with an issue? Are there any important gaps or missing levels?	Do these levels work together and do they trust each other between levels? To what degree is the mutual dependence among levels recognised?	Is it possible to move up and down levels (upscaling and downscaling) given the issue at stake?	Is there a strong impact from a certain level towards behavioural change or management reform?
Actors and networks	Are all relevant stakeholders involved? Are there any stakeholders not involved or even excluded?	What is the strength of interactions between stakeholders? In what ways are these interactions institutionalised in stable structures? Do the stakeholders have experience in working together? Do they trust and respect each other?	Is it possible that new actors are included or even that the lead shifts from one actor to another when there are pragmatic reasons for this? Do the actors share in 'social capital' allowing them to support each other's tasks?	Is there a strong pressure from an actor or actor coalition towards behavioural change or management reform?
Problem perspectives and goal ambitions	To what extent are the various problem perspectives taken into account?	To what extent do the various perspectives and goals support each other, or are they in competition or conflict?	Are there opportunities to re-assess goals? Can multiple goals be optimized in package deals?	How different are the goal ambitions from the status quo or business as usual?
Strategies and instruments	What types of instruments are included in the policy strategy? Are there any excluded types? Are monitoring and enforcement instruments included?	To what extent is the incentive system based on synergy? Are trade-offs in cost benefits and distributional effects considered? Are there any overlaps or conflicts of incentives created by the included policy instruments?	Are there opportunities to combine or make use of different types of instruments? Is there a choice?	What is the implied behavioural deviation from current practice and how strongly do the instruments require and enforce this?
Responsi-bilities and resources	Are all responsibilities clearly assigned and facilitated with resources?	To what extent do the assigned responsibilities create competence struggles or cooperation within or across institutions? Are they considered legitimate by the main stakeholders?	To what extent is it possible to pool the assigned responsibilities and resources as long as accountability and transparency are not compromised?	Is the amount of allocated resources sufficient to implement the measures needed for the intended change?

Fig. 3.6 The governance assessment tool matrix with its main evaluative questions (*Source* Bressers et al. 2013)

3.4 Using the Governance Assessment Tool

It is important to note that even with all the questions that specify the cells of the matrix, hard "measurement" in the sense of a quantification is not possible. Some degree of "informed judgment" is inevitable when assessing the status of the four criteria relevant to the various governance dimensions.

The GAT can be used by stakeholders themselves, or as a guidance for inter-active workshops with stakeholders. In the DROP project we had the opportunity to use the GAT in a very elaborated way. Thus it makes sense to first explain some options on how to use the GAT in a situation that time and involvement are limited. Thereafter we will explain how we used the tool in the DROP project and what we recognised to be the main success factors.

3.4.1 Diagnosing with the Governance Assessment Tool in a Short Period and with a Limited Number of People

The structure of the Governance Assessment Tool and the guiding questions it poses in each cell, enables any *individual stakeholder* (e.g. a project leader, or policy advisor, or policy makers) who understands the dimensions and criteria to assess the governance context s/he is working in. All it requires is a few hours to assess the situation for each cell, on the basis of knowledge held by heart. Obviously such assessment is limited by the degree of correctness of such esti-mates. But that is no reason to be negative about it. Such an individual thought experiment at least turns implicit knowledge and perceptions into explicit ones that can later be shared with others in a systematic way. It also serves the purpose that the individual stakeholder becomes more aware of the issues on which there is uncertainty or even lack of knowledge. Lastly, assuming that the perceptions of such 'insider' make some sense indeed (as often will be the case), it provides an assessment of one's own working circumstances that can be practical in finding ways to improve them or otherwise deal with them.

A next step in elaborated use of the tool is when a *group of practitioners interactively* uses it for a systematic brainstorm on the governance context of their common policy or project. This could take for instance the form of a half day workshop. Compared with the previous approach there are more people that can contribute knowledge and that can counter one-sided bias in perceptions, creating a degree of "inter-subjectivity". The joint effort is also an important aspect in itself, as it provides a basis for sharing information and sharing perceptions, that can later be of utmost value for productive collaboration (HarmoniCOP 2005). The session can be concluded by brainstorming on how to deal with the governance context about which by then a joint understanding has evolved. In as far as differences of opinion occur and persist, the session has probably pinpointed more precisely than before where the disagreement is all about.

A variant of the above is the situation in which an experienced analyst, for instance a scientist that worked with the tool more often, leads the session, turning it into a *guided workshop*. An obvious advantage is that the governance expert has a good understanding of the precise meaning of the concepts and the reasons why they are included in a model explaining the degree to which the context is supportive or not. This can help the participants to concentrate on the substantive matters while nevertheless all cells of the assessment tool get appropriate attention. The governance expert can also help in the interpretation of the consequences of the assessment and will develop experience in how to deal with such situations, creating learning from one case to another. The disadvantage however can be that the participants are less actively involved and feel more like interviewees than like discussants. A good balance between too much and too little guidance is important for a productive process that provides the participants with learning experience.

A further way in which the tool can be applied is when not practical learning experiences of practitioners, but *scientific research* is the main purpose. In such a project all kind of sources are used to assess the cells and interviews with practitioners are just part of the data gathering. The Governance Assessment Tool will in such studies often be used as a way to "measure" the dependent or independent variable in the study. For this purpose normal approaches to methodology apply.

One more way in which the tool can be applied is the *elaborate way it is done in the DROP project* (multiple analysts from multiple institutions, disciplines and countries, in multiple rounds and various ways of interaction with practitioners). This is a very special situation that requires much resources, but provided both scientifically and practically a lot of new knowledge. About this ideal methodology (and its risks and pitfalls!) the next subsection will elaborate further.

3.4.2 Diagnosing with the Governance Assessment Tool in the DROP Project

In this section both experiences with using the GAT in the DROP project will be shared and advices for potential users will be presented that are based on the lessons learned while using the GAT. While the text contains a lot of advices on how to use the GAT, it does not have the character of a manual. As regards to the implementation of the GAT in the case of the DROP project, a number of important factors can be highlighted explaining the relative success of the project. Many of them relate to the challenge of using the tool to assess a variety of cases with different main policies and projects in various national and regional conditions. This requires both a good common understanding of the concepts in the GAT and a good common understanding of the empirical situation that can be supported by the items listed below.

Continuous iteration between science and practice—A way to ensure the valid and reliable assessment of the governance context of a particular region is through liaising with those embedded strongly in the governance context and water management reality. In the DROP project, the governance assessment has been developed by social scientists with the help of the practice partners (project partners from the region such as water authorities and county councils) and other governmental and non-governmental stakeholders. This has allowed both for continuous iteration between science and practice, as well as for access to regional stakeholders for interviews to ensure an even representation of relevant stakeholders. In order to enable a complete coverage of the perspectives and opinions of different stake- holders, the governance team visited each region twice and prepared a draft assessment report for each region, which was finalised after the second round of visits. The practice partners and other stakeholders interviewed were encouraged to 'feedback' into the draft reports to ensure that the governance assessment reflected the reality of water management in that specific regional context. Having exchanges with practice partners on the governance assessment can also contribute substan- tially to the development of recommendations. It is relatively easy to propose that some action should be undertaken to improve or circumvent weaknesses in the governance context. But the development of advices about how to implement such actions needs inputs from the practice partners.

Diversity in backgrounds of the analysts diagnosing the governance context played a positive role. It helps avoiding scientific disciplinary terminology. Simple and clear messages are easier to translate into concrete and feasible actions. In an interdisciplinary project team, there is a constant need for mutual adjustment and searching for a common language. Equally, that the governance team was com- posed of 'outsiders' to the region meant that there was objective reflection on what were sometimes very local issues and to the institutional rules and habits involved. Questioning what would otherwise be taken for granted by observers from the own country or region, can provide important eye-opening reflections.

Visit several regions—In DROP we found it quite useful to have the team members visit several regions. This observation of several governance settings allows for comparative analysis already during the data gathering phase, and as a result that creates the possibility to sharpen questions along the way. Most members of the governance team visited two or three regions twice, one team member even visited all regions.

Awareness raising 'intervention'—Doing such research in the region also forms a type of awareness raising 'intervention'. A number of the stakeholders inter- viewed had a fairly low awareness of the relevance of drought for their cases and the role of climate change therein. Nevertheless, the fact that an international governance team was visiting their region, asking many questions on the subject and returning with feedback and further questions half a year later contributed to pushing drought and water scarcity onto regional agendas. The modification side-effect of such 'action research' can also inhibit local actors in participating fully in the assessment. The fact that the GAT is not meant to evaluate the work of the practice partners, but the context under which they have to do their work should be made clear and might help in this respect.

Pre-collect existing documentation—Given the diversity of nationalities and diversity of professional backgrounds of the governance assessment team, it was found to be useful to collect some existing documentation or prepare a short document to provide prior information on the context of climate change, water management, and other relevant policies of each region. This levelled up the governance team members' understanding of the main features of each site before the interviews. It also allowed the interviews to focus on issues that were not published or available elsewhere, thus using the short time of the interviews more efficiently.

Governance analysts translating themselves—When translation was needed between the English language questions of the governance team members and the representatives of stakeholders that were not comfortable in that language, it proved to be good to have one of the governance team members to fulfil the "translator" role. This way the relation between interviewers and interviewees did not get disrupted and the knowledge of the tool by the governance team member ensured good interpretations and summaries of what was said by the stakeholders. Furthermore it proved to be particularly useful to be able to adapt the questions to the case by using terms of local institutions.

Local institution contacting stakeholders—The inclusion of a local institution as a cooperative partner for the interviews was very useful for contacting relevant stakeholders. The local partner possessed a well-established network and could more easily convince stakeholders to participate in interviews. Additionally, the local partner was central in compiling and screening the most relevant stakeholders and actors, including less obvious groups, to interview for achieving the widest scope possible. The assessment team made sure that also potential critics were involved among the stakeholders interviewed. A problem occurs if a major stake-holder group cannot be reached, because then the point of view of this group cannot be involved in the discussions. During the second visits, the governance team tried to make up for such situations, in some cases by visiting those stakeholders at their own offices or even homes. Like with all evaluations, selected outcomes of the governance analysis can also be used by actors as a tool in power relations. While the use of the GAT requires mutual trust between the interviewed actors and the governance team, a neutral position is required, as well as a capability to understand and integrate various positions.

The interviews had a variety of settings—Some were individual interviews and some group interviews to test the efficiency of each approach. The analysis is very much dependent on open discussions between the interviewees and the inter-viewers. It is necessary to gather critical issues, therefore individual interviews or small groups of interviewees seemed more suitable for the establishment of trust, and the open discussion of sometimes critical or difficult issues. Another experience was that the presence of a representative of the practice partner (water authority) at the interviews was sometimes useful to get a good introduction of the governance assessment exercise to the interviewees, but should also be dealt with carefully in order to make sure that the interviewees feel they can talk freely.

Generally the GAT should not be used as a battery of questions to put forward during each interview, but *used as a checklist* to make sure that all issues were dealt with in the course of the conversation while keeping the flow of the conversation as much as possible. The questions from the GAT should be adapted to the local contours of each case, such that the questions targeted the specific local context, including appropriate strategies and instruments, local actors, and level of analysis.

Debriefing sessions—Through a series of short debriefing sessions by the governance assessment team directly after each round of interviews, the data was extracted and analysed in the context of the 20 evaluation items of the GAT matrix. Within a week after each session, a teleconference (phone or skype meeting) provided additional exchange and inputs to the main authors of each case report. The draft case reports were distributed for comments among the governance team members, and ultimately discussed with the practice partners during the second visit to the case areas. These draft reports also formed the basis for judging what issues to focus on in the second round of interviews—as the development of these reports allowed the identification of issues or stakeholder perspectives missing in the assessment.

Careful summarising of results—The results of a GAT analysis can be summarised, even in the form of figures or tables. The issue is that transferring the richness of the data gathered by numerous documents and interviews into more condensed layers of summary and ultimately into an overview has both positive and negative aspects. On the one hand it is necessary to enable comparative analysis between several cases. On the other hand, the summary should not hide away essential observations that form the evidence for the scores. In the DROP project this has been achieved by assessing each of the twenty cells of the matrix by a brief statement and sometimes a score at a three or five point scale, followed by a paragraph to page length of observations on which this statement is based. The scores on a three point scale have also been translated into graphical visualisations showing the matrix with colours ('score cards') indicating the value of each cell. These visualisations enable a quick overview of the results. However, one should always keep in mind that such a summary of summary is a derivate of a much richer set of observations and its interpretation.

Comparative analysis—The multiple case study character of the use of the GAT helped us to develop recommendations based on what works well elsewhere and what stands out in one region compared to other regions. Insights from pilot cases that face similar challenges are potential sources of advices, with the benefit of having a clear example to illustrate the ideas with concrete outcomes. Additionally, as a contribution for the learning experience of the practice partners, hearing about the governance assessment conclusions regarding other regions provides the possibility to refresh the way their own context is reviewed.

Procedure to compile recommendations—Statements of the different cells and questions of the assessment matrix were screened carefully. Important connecting issues were then highlighted. Especially the critical statements, which the stakeholders made during the interviews, were screened by the governance team to identify the improvement areas. This brainstorming exchange within the governance team was useful in developing and structuring ideas relevant to the recommendations. Comparisons between the different case studies were also explored to identify common issues as well as opportunities among the case studies. Different approaches and experiences could be compared and used as the basis for further discussion. One major step was to gather feedback to the developed recommendations. It was evident that the recommendations were developed with limited knowledge of the history of the local and regional institutions and their culture and experiences. As a result, it was very important to discuss the recommendations and gather feedback.

3.5 Summary and Conclusion

In this chapter we have introduced the Governance Assessment Tool that has been used in the DROP project and forms the analytical basis of this book. We started with the origins of the tool in Contextual Interaction Theory, and proceeded with the dimensions and criteria that form the backbone of the tool, and form a matrix. In these matrix evaluative questions are formulated that can be discussed with local and regional stakeholders. Based on their answers and further information and insights a judgment can be reached to what extent the governance circumstances are supportive, restrictive or neutral for drought adaptation. A visualisation with coloured cells of the matrix can show in one quick glance the governance state of affairs in that region. To create more precise visualisation, arrows can be added to each box indicating upward or downward trends for that box. The chapter ends with a discussion on the application of the GAT. The tool can both be used in relatively simple ways and as in the DROP project in a very elaborate way.

References

Axelrod R (ed) (1976) Structure of decision. The cognitive maps of political elites. Princeton University Press, Princeton
Bandura A (1986) Social foundations of thought and action. A social cognitive theory. Prentice Hall, Englewood Cliffs
Bressers H (2004) Implementing sustainable development: how to know what works where when and how. In: Lafferty WM (ed) Governance for sustainable development: the challenge of adapting form to function. Edward Elgar, Cheltenham Northampton, pp 284–318
Bressers H (2009) From public administration to policy networks. Contextual interaction analysis. In: Stéphane N, Varone F (eds) Rediscovering public law and public administration in comparative policy analysis. A tribute to Peter Knoepfel. Presses polytechniques, Lausanne, pp 123–142

Bressers H, Klok PJ (1988) Fundamental for a theory of policy instruments. Int J Soc Econ 15 (3/4):22–41

Bressers H, Kuks S (2003) What does "governance" mean? From conception to elaboration. In: Bressers H, Rosenbaum W (eds) Achieving sustainable development: the challenge of governance across social scales. Praeger, Westport Connecticut, pp 65–88

Bressers H, Kuks S (eds) (2004) Integrated governance and water basin management. Conditions for regime change and sustainability. Kluwer Academic Publishers, Dordrecht

Bressers H, Lulofs K (2009) Explaining the impact of the 1991 and the 2000 firework blasts in the Netherlands by the core of five policy change models. In: Capano G, Howlett M (eds) European and North American policy change. Drivers and dynamics. Routledge/ECPR, New York, pp 15–42

Bressers H, Lulofs K (eds) (2010) Governance and complexity in water management. Creating cooperation through boundary spanning strategies. Edward Elgar, Cheltenham

Bressers H, O'Toole LJ Jr (2005) Instrument selection and implementation in a networked context. In: Elidas P, Hill MM, Howlett M (eds) Designing government. From instruments to governance. McGill-Queen's University, Montreal, pp 132–153

Bressers H, de Boer C, Lordkipanidze M, Özerol G, Vinke-De Kruijf J, Furusho C, La Jeunesse I, Larrue C, Ramos MH, Kampa E, Stein U, Tröltzsch J, Vidaurre R, Browne A (2013) Water governance assessment tool. With an elaboration for drought resilience. Report to the DROP project, CSTM University of Twente, Enschede

Bressers N et al (2015) Practice measures example book DROP: a handbook for regional water authorities. Water authority Vechtstromen, Almelo

de Boer C, Bressers H (2011) Complex and dynamic implementation processes. Analyzing the renaturalization of the Dutch Regge river. University of Twente and Water Governance Centre, Enschede

Dryzek JS (1997) The politics of the earth. Environmental Discourses. Oxford University Press, Oxford

Edelenbos J, Bressers N, Scholten P (2013) Water governance as connective capacity. Ashgate Publishing, Farnham

Fischer F (1995) Evaluating public policy. Nelson Hall, Chicago

Fischer F, Gottweis H (eds) (2012) The argumentative turn revisited public policy as communicative practice. Duke University Press, Durham

Gatersleben B, Vlek C (1998) Household consumption quality of life and environmental impacts: a psychological perspective and empirical study. In: Noorman KJ, Uiterkamp TS (eds) Green households? Domestic consumers, environment and sustainability. Earthscan, London, pp 141–183

Gerrits L (2008) The gentle art of coevolution. EUR, Rotterdam

Goggin ML (1986) The 'too few cases/too many variables' problem in implementation research. Polit Res Q 39(2):328–347

HarmoniCOP (2005) Learning together to manage together: Improving participation in water management. University of Osnabrück, Osnabrück

Howlett M (2011) Designing public policies. Principles and instruments. Routledge, London

Klok PJ (1995) A classification of instruments for environmental policy. In: Dente B (ed) Environmental policy in search of new instruments. Kluwer academic, Dordrecht, pp 21–36

Knoepfel P, Kissling-Näf I, Varone F (2001) Institutionelle regime für natürliche ressourcen. Helbing & Lichtenhahn, Basel

Knoepfel P, Kissling-Näf I, Varone F (eds) (2003) Institutionelle Ressourcenregime in Aktion/ Régimes institutionnels de ressources naturelles en action. Helbing & Lichtenhahn, Basel

Knoepfel P, Larrue C, Varone F, Hill M (eds) (2011) Public policy analysis. The Policy Press, Bristol

Kuks S, Bressers H, Boer C de, Vinke J, Özerol G (2012) Governance assessment tool. Institutional capacity. Report to the Water Governance Centre, University of Twente, Enschede

Mohlakoana N (2014) Implementing the South African free basic alternative energy policy: a dynamic actor interaction. University of Twente, Enschede

Owens K (2008) Understanding how actors influence policy implementation. A comparative study of wetland restorations in New Jersey, Oregon, The Netherlands and Finland. University of Twente, Enschede

Sabatier PA, Jenkins-Smith HC (1999) The advocacy coalition framework: an assessment. In: Sabatier PA (ed) Theories of the policy process. Westview Press, Boulder, pp 117–168

Schön DA (1983) The reflective practitioner. How professionals think in action. Basic Books, New York

Schön DA, Rein M (1994) Frame reflection toward the resolution of intractable policy controversies (Chap. 247). Basic Books, New York

Teisman G, Gerrits L, van Buuren A (2009) An introduction to understanding and managing complex process systems. In: Teisman G, Gerrits L, van Buuren A (eds) Managing complex governance systems. Routledge, New York, pp 1–16

True JL, Jones BD, Baumgartner FR (1999) Punctuated equilibrium theory: explaining stability and change in American policy making. In: Sabatier PA (ed) Theories of the policy process. Westview Press, Boulder, pp 97–115

van Hulst M, Yanow D (2014) From policy "frames" to "framing". Theorizing a more dynamic political approach. Am Rev Publ Adm 1–21

Chapter 4
Eifel-Rur: Old Water Rights and Fixed Frameworks for Action

Rodrigo Vidaurre, Ulf Stein, Alison Browne, Maia Lordkipanidze,
Carina Furusho, Antje Goedeking, Herbert Polczyk
and Christof Homann

4.1 Introduction

This chapter summarises our analysis of drought governance in the Eifel-Rur region
of Germany. Within the Interreg IV-B project DROP a team of researchers from five
universities and knowledge institutes performed two field visits to the Eifel-Rur
region and held interviews with authorities and stakeholders. The visits were facilitated by the DROP project partner Eifel-Rur Waterboard (Wasserverband Eifel-Rur,
WVER). Interviews were both individual and in group settings; in the second visit
interim results were presented to stakeholders in a workshop. Stakeholders interviewed included representatives from drinking water producers, nature protection
authorities, industrial water users, farmer representatives, electricity generating
companies, environmental NGOs, fishermen, sailing schools, and local (municipal)
and regional (district) authorities in charge of water management. The analysis was
guided by a drought-specific Governance Assessment Tool (GAT), which uses five
governance dimensions (levels and scales, actors and networks, problem perceptions
and goal ambitions, strategies and instruments, responsibilities and resources) and
four governance criteria (extent, coherence, flexibility and intensity) in its analysis.

R. Vidaurre (✉) · U. Stein
Ecologic Institute, Pfalzburger Strasse 43/44, 10717 Berlin, Germany
e-mail: rodrigo.vidaurre@ecologic.eu

A. Browne
University of Manchester, Manchester M13 9PL, UK

M. Lordkipanidze
University of Twente, PO Box 7658, 8903 JR Leeuwarden, The Netherlands

C. Furusho
IRSTEA, Antony, France

A. Goedeking · H. Polczyk · C. Homann
Eifel-Rur Waterboard—WVER, Düren, Germany

© The Author(s) 2016
H. Bressers et al. (eds.), *Governance for Drought Resilience*,
DOI 10.1007/978-3-319-29671-5_4

In the following, we present the context of water management in the Eifel-Rur region, describe some drought actions which have already been implemented, explain the results of our analysis in terms of the Governance Assessment Toolkit and present our recommendations for improved drought governance in the region.

An aspect of the Eifel-Rur water management system which is central for its drought governance is the water rights' system in place. The region's water rights—some of them centuries old—provide their owners with very strong legal claims to the resource; furthermore, the current system of rights and charges does not provide real incentives for users to reduce their water rights/water use. These features make the demand side of the water system very inflexible. In addition, the region's significant number of reservoirs allows for a very stable supply of water over time, which means that users are not prone to include risks related to water supply into their risk strategies. This lack of flexibility poses significant challenges for drought management, some of which are taken up in the final section "Conclusions and Case-Specific Recommendations".

4.2   The Who, What and When of Drought Governance in the Eifel-Rur Region

4.2.1   Water Management in North Rhine-Westphalia

In Germany, the EU's Water Framework Directive (WFD) was transposed into national law via the seventh amendment to the Federal Water Act (*Wasserhaushaltsgesetz*, WHG) in June 2002.[1] Due to a major restructuring of responsibilities and competencies between the Federal Government and the German *Länder* in 2006, the German water legislation was modified in 2009; the modified Federal Water Act entered into force in 2010.

According to this act, in their implementation of the WFD the German federal states must adopt their state water laws to encompass water protection and to formulate the roles for cities, municipalities and water authorities, who bear the concrete responsibility for implementing measures. In the case of North Rhine-Westphalia, the responsibility for developing the river basin management plans lies with the Highest Level Water Authority which is the North Rhine-Westphalian Ministry for Environment and Nature Protection, Agriculture and Consumer Protection. Plans are adopted in consultation with the High Water Level Authorities (District Councils) and the responsible committee of the North Rhine-Westphalia regional parliament (Landtag). Responsibility for implementation lies with lower level public administration, such as districts and cities. Further actors such as nature protection organisations, water associations and regional councils should participate in the planning and particularly in the implementation process. Regarding water abstractions, it is the

---

[1]Grüne Liga (n.d.): Umsetzung der Richtlinie in deutsches Recht. http://www.wrrl-info.de/docs/tafel7_a3.pdf.

District Councils who are responsible for authorisation of water abstraction for surface water and groundwater.

In the particular case of the Eifel-Rur river basin, the district government in Cologne (Aachen) is responsible for implementation of the WFD on the ground. The measures are financed 80 % from the state government and 20 % from own contribution (e.g. the municipalities where they are responsible).

North Rhine-Westphalia's water management is quite particular in the German context, as it relies on waterboards to perform many of the duties of water management. This particular form of organising water management has its origins in the nineteenth century, in response to the large-scale water-related challenges of North Rhine-Westphalian coal mining. The responsibilities of the waterboards are established in a particular law for each single waterboard. The next section describes the responsibilities of the WVER.

4.2.2 The Eifel-Rur Waterboard (WVER)

WVER is a public water corporation in the district of Cologne (one of the five governmental districts of North Rhine-Westphalia), similar in nature to a water authority. It is a public body which is an operating organisation, executing different tasks set by the special North Rhine-Westphalian law *Gesetz über den Wasserverband Eifel-Rur* ("Law on the Water Association Eifel-Rur"). An important point is that the WVER is limited to executing powers, without any rights of an authority (e.g. it cannot issue permits). The WVER region comprises mainly the catchment of the Rur and has approx. 2.087 km^2 and ca. 1.1 million inhabitants.

WVER responsibilities comprise the full range of water services. Duties of WVER by law include control of water discharge in catchment area, river maintenance, river restoration, supply of raw water for drinking water production, supply of production water, wastewater treatment, prevention of disadvantageous influences on river systems (in general looking at different issues) and hydrology. Groundwater is not included under WVER's duties, as only the northern low-lying part of WVER's area has significant groundwater bodies. In this region open-cast coal mines are situated which influence the groundwater table, but even larger mines are situated in the adjacent catchment area, which also influence the groundwater table in the northern Rur region. As a consequence, groundwater management has been entrusted to the waterboard in this neighbouring catchment.

In addition to its legal obligations, WVER informally collaborates with further actors to achieve additional objectives including keeping reservoir levels high enough for water quality, sailing and to ensure a pleasant landscape (tourism); managing reservoir levels in a way that minimises disturbances of fish reproduction, and electricity production by the company RWE.

WVER operates six reservoirs with a total capacity of 300 million cubic metres in the northern part of the Eifel hills, which corresponds to the southern part of its service area. The reservoirs were mainly developed for flood control and flow maintenance during dry seasons. Stillwater in these reservoirs always bears the risk of eutrophication with effects such as algal blooms, etc. Concerning climate change, with longer dry and sunny periods, this problem is expected to increase.

The total length of flowing surface waters in the WVER service area managed by the waterboard is approx. 1900 km. (These are all the waters in the northern part of the service area downstream of the reservoirs.) WVER is responsible for the management of these waters, as well as for the operation of 50 flood retention basins and other flood control works.

4.2.3   The Role of Municipalities and Lower Water Authorities in Water Management

The German Basic Law (Article 28 (2)) and most constitutions of the German *Länder* ensure the local self-government of districts, towns and municipalities. Self-government comprises all matters concerning the local community. Municipal regulations and the water laws of the different German federal states stipulate that drinking water supply is usually, and wastewater disposal is always, an obligation of the local authorities. On that basis, municipalities decide on the local implementation and organisation of water supply and wastewater disposal.

With a view to effectively realising drinking water supply and wastewater disposal, municipalities may form associations for voluntary cooperation. To some extent, municipalities (such as in North Rhine-Westphalia) are members of water management associations (*Wasserverbände*, such as the Waterboard Eifel-Rur), which are subject to special laws.[2] In addition to these compulsory tasks, municipalities have to fulfil partial tasks regarding the implementation of environmental laws issued by the government and the German Federal States.

Among other responsibilities, the lower water authorities, as supervisory/executive authorities, approve flooding areas, wastewater systems, wastewater treatment plants, small sewage works, wastewater and rainwater discharges, water supply facilities, the use of water bodies, such as abstraction from surface water and exceptional approvals for water and medicinal spring protection areas.

4.2.4   Historical Approach to Droughts and Their Effects on Drinking Water and Water Quality

There is as yet no strategic, long-term approach to drought management in North Rhine-Westphalia's water management. There is also no incorporation of climate

---

[2]Profile of the German Water Sector (2011), available under http://www.dvgw.de/fileadmin/dvgw/wasser/organisation/branchenbild2011_en.pdf.

change and its impacts on water availability in the planning tools and instruments used in water management of the Waterboard Eifel-Rur. Water management is based on historical data sets, and no prognoses have been developed to account for altered conditions in the future. At the moment of writing, and excepting the work performed in the DROP pilot (see next section), only a first prognosis on water quantity in the Rur system within different climate change scenarios has been developed (within the AMICE project, a further INTERREG IV-B project). A prognosis on water quality is to the moment also lacking.[3]

North Rhine-Westphalia has as yet not much experience with drought episodes; it has a comparatively humid climate due to its proximity to the Atlantic. In the case of the Waterboard Eifel-Rur, the few droughts in the past have been dealt with on an ad hoc basis: water management measures have been developed that alleviate a particular impact over a short period of time. The requirement for action has arisen not due to considerations related to droughts themselves (e.g. anticipatory management in early stages of drought to prepare for possible worsening), but due to other requirements, such as upholding water quality commitments.

4.3 Measures Taken: Addressing Drought in the Eifel

In the upper catchment of the Rur, six reservoirs were built in the Eifel hills mainly to control the effects of flooding and to maintain the flow during dry seasons. Five of them form an interconnected system around the main reservoir 'Rurtalsperre'. The most upstream dam is placed in the tributary Olef and called 'Oleftalsperre'. It is a multifunction reservoir with a storage capacity of 19 mio. m3. The 'Oleftalsperre' was built for the protection against floods, for low-water enrichment and for the provision of raw water for tap-water production. The Olef mouths info the next tributary the Urft. There the 'Urfttalsperre' is situated. It is the oldest dam in the northern Eifel with a storage capacity of 45 mio. m3. The outflow of the dam flows in very dry periods directly into the next basin 'Obersee', a preimpoundment basin of the 'Rurtalsperre'. This next basin serves among other things also as a reservoir providing the agency in charge with water for production of drinking water. The 'Obersee' flows into the biggest reservoir the 'Rurtalsperre' (202 mio. m3). This is also a multifunction reservoir without direct storage for tap-water production, but among other things a lot of tourism, which is based on a large lake with good water quality.

These reservoirs were often shaped as filled constructions with a stream/river flowing through the basin as a stream. Big reservoirs in the catchment area of the Rur such as the dams in the northern Eifel cannot be disconnected from the river, because their retention volume cannot be replaced by a near-natural reconstruction of the river course. With a total capacity of 300 mio. m3 their function for flood

---

[3]Antje Goedeking, WVER, personal communication.

control is very important. Consequently, an adaptation to climate change is only possible by an adaptation of the management of the dams.

In addition to flood protection, the 'Rurtalsperre' reservoir system serves additional important aims. Among the reservoir system's functions is that of providing good quality raw water for drinking water production. Whereas the different aims do not always go in line with each other, still all of them have to be served. For example, sometimes a controlled high discharge out of the reservoir is needed in order to prevent flooding, but this can only be carried out to such an extent that there is still enough water in the reservoirs to produce drinking water and maintain the flow in dry periods.

Recently, Eifel-Rur region has experienced somewhat dryer periods during the spring season, as a result of which the water flow through the reservoirs decreases. Stillwater and falling water levels in reservoirs bear the risk of a decrease in water quality, which results in a higher amount of production work and possibly drinking water production problems; stillwater in these reservoirs always bears the risk of eutrophication with effects such as algal blooms. Due to the topography and the limited capacity, the 'Oleftalsperre' and the 'Urfttalsperre' run the risk of more algal blooms during long dry periods. This can also have consequences on the reservoirs 'Obersee' and 'Rurtalsperre' downstream.

Concerning climate change with longer dry and sunny periods this problem is expected to increase. At present there is a lack of knowledge about the behaviour of the water quality within the dry scenario. The long dry periods in spring in the last years already resulted in a loss of water quality in some of the reservoirs.

The following map presents the main surface water reservoirs' location (Fig. 4.1).

The Waterboard Eifel-Rur has executed a project to improve water reservoir management. The project aims to prevent deterioration of the water quality in the reservoir system. To this purpose the waterboard analysed the inflow patterns in the different dams. Based on these results, a study was carried out on the management system of the dams with respect to water quantity and quality. Suggestions for the adaptation of the management plan emerged: one of the best results obtained is to add a drought index in the management plan, which will help prevent the release of too much discharge in an earlier stage compared to today's practice. This leads to a credit of water in dry periods.

The project thus managed to flexibilise operational decisions to improve the performance of the management system, in which the different obligations of the system are now still met under a wider array of meteorological and flow conditions. Whereas certain dry conditions in the past would have made it impossible to meet all obligations, under the improved system this would now be possible. However, the issue of *flexibilising the obligations* is in our opinion not yet satisfactorily addressed. Particularly, the water rights regime ensures constant supply to water users, and contains no incentives to reduce these water rights where there could be potential for such reductions. The following sections present this situation in more detail.

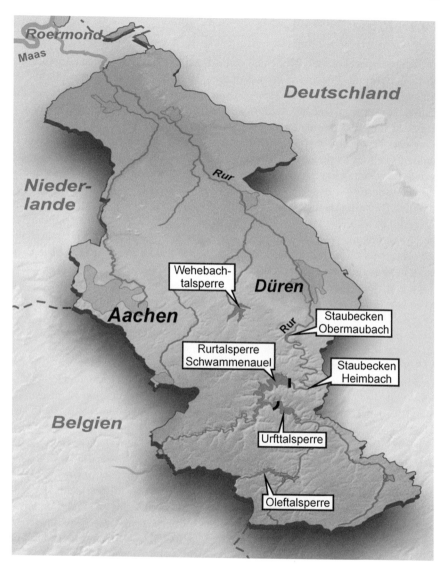

Fig. 4.1 WVER region in the catchment of the Rur River, including the the main reservoirs in the Eifel region

4.4   Governance Assessment: From High Coherence to Low Flexibility

In the following section an analysis of drought governance in the Eifel-Rur region is presented. It is structured along the four qualities of the GAT.

4.4.1 Extent

The *extent* aspects of the governance context can mostly be regarded as somewhat positive, covering all levels and scales of the system. Many administrative levels are directly involved in the water management system. The two main actors are the district government (second authority level) and the waterboard itself. When it comes to droughts, the national level (German Ministry of Environment) is still somewhat decoupled, mainly providing first studies and visions. The EU level is quite relevant for its directives. However, a negative point is that municipalities are seen to be withdrawing from their water management responsibilities, mainly due to serious resource issues.

When focussing on the actors, the same positive extent can be appreciated. This is a result of the design of the North Rhine-Westphalian waterboards: according to the law regulating WVER, users with a water right of a certain size are automatically members of the waterboard, whether they like it or not. This means that all major users participate—also economically—in the management of the water basin. There are, however, some restrictions to this positive extent regarding actor involvement. Smaller (and thus non-paying) actors, such as farmers and nature organisations, do not have the same voice as larger actors. This been said, there is a strong movement towards collaborative and inclusive decision-making processes on the side of the water authorities, as well as on relationship building on the side of WVER. The implementation of the WFD and the Floods Directive for instance were based on a huge number of participatory workshops and roundtable discussions, and there is a strong emphasis on voluntary implementation of measures. Interviewed stakeholders repeatedly mentioned that the developments over the last decades had been very positive in this sense.

The implementation of WFD and Floods Directive has provided the region with a set of new instruments and experience in consultation processes with stakeholders, and all in all, there is a broad extent of strategies and instruments in place. However, from a *drought* perspective there are very significant elements still missing, e.g. water demand management, drought contingency planning, communication, etc. In this context there is still room for improvement, e.g. via knowledge transfer of experiences from other pilot regions. Another point affecting drought management in particular is the fact that the district government does not currently see itself in a position to actively push the topic of droughts—whereas they welcome the waterboard's actions on the topic, they are currently suffering due to overstretched resources. This means that those actors could implement measures on the ground if they were required to face no external pressure to act on the topic.

4.4.2 *Coherence*

Passing to the governance system's *coherence*, the evaluation is also quite positive. The main actors, such as the state level, the districts, the municipalities, the water authorities and the drinking water companies, mutually accept their share of the tasks, responsibilities and funding given by law. The dependence among these levels is well recognised by the interviewed individuals. EU environmental policies seem to have played an important role in introducing a more holistic and synergistic approach to the management of the reservoirs. To some degree the coordination of the lower competent authorities appears to be more coherent than that at a higher level.

Among the factors determining a positive degree of coherence is that the WVER is in charge of practically all relevant water management tasks in the region. All these responsibilities being within one organisation rather than distributed between different actors is probably helpful in establishing a coherent framework. The institutional structure of WVER also helps that with water users also being the waterboard members, involved in decision-making and paying for the services provided, this structure provides a framework conducive to good coherence of, for example, perceptions, goal ambitions, strategies and instruments. In addition, different stakeholders have goals that match quite well. For instance, the fishermen associations are interested in large fish populations in the reservoirs, which are also of interest to the waterboard in their role of drinking water supplier because of fish population's positive effects on water quality, in their role of responsible for WFD implementation, and also for the objectives of the national park authorities.

In the WVER region, the interviews show a sense of trust and mutual dependency between the actors, expressed for instance in their positive evaluation of participatory approaches being used in water management. All actors interviewed were quite satisfied by the way the waterboard is working with them and how actors' perspectives are considered when proposing measures, for instance, for the implementation of the WFD. All in all, the stakeholders interviewed expressed their belief in the extremely high value of the trust-based collaboration that has been built over the years, and that has evolved positively over time. However, the consensual and voluntary approach towards measure implementation seems in some cases to be reaching its limits, with some negotiation processes on contentious topics being practically at a standstill for a number of years now.

This notwithstanding, we can identify potential conflicts of interest that could worsen in case of water scarcity. The existence of very old water rights (with strong legal precedence) seems to create opposing goals between some users from a drought perspective. Industry users with a certain water right do not at present have incentives to reduce their water use or to partly reduce their water rights. A further point is that the strategy for flood prevention implies keeping the water level in the reservoir sufficiently low until the spring, to ensure enough storage capacity in case of exceptional precipitation events which may be associated with intense rainfall or snow melt. However, if there is not enough precipitation or snow melt during the spring period when water is collected, there is not enough water meeting the quality conditions for some drinking water providers (e.g. water temperature below 10 °C and oxygen above 4 mg/l). Furthermore, there is a lack of coherence when it comes to resources; in particular, there are a lot of issues with municipalities being extremely cash-strapped at the moment as well as in the foreseeable future.

This said, it is also true that drought can be considered a second-order problem and the extent to which conflicts related to drought and water scarcity have emerged is really quite limited—with the exception of punctual issues of water supply and water quality between core actors WVER and a water supply company.

4.4.3 Flexibility

The overall evaluation of the governance system's *flexibility* is only intermediate with, however, some positive developments over time. This evaluation is based on the fact that the water management system has a quite rigid large-scale framework, shown fundamentally in the priorities and responsibilities of WVER (established by law) and its operational procedures. The management of the water system in the Eifel valley follows a clearly established set of complex management rules which WVER helps elaborate and which are authorised by the district authority of Cologne. Any management decision that disregards these rules can bring with it the question of legal responsibility—for instance for flood damages ensuing due to incorrect flood protection. This means that WVER and its personnel have a strong incentive not to stray from this set of rules.

The framework is both difficult and slow to change, and some actors see it as problematic for the system to take on-board new responsibilities. However, the water management system shows very significant flexibility at the small scale, within the rooms provided by this overall fixed framework. There is a strong culture of discussion and collaboration between actors, and interviewed stakeholders were broadly of the opinion that their interests are considered and taken on board as much as possible.

The legal obligation of the waterboard to provide a certain established level of protection (floods) and of supply (deliver water for drinking water production) and the responsibilities associated with it have resulted in an elaborate and sophisticated set of rules that manage the interaction of reservoirs and water bodies. However, these same legal obligations imply that there is no short-term possibility of officially incorporating additional risks (e.g. droughts) into the set of principles which govern the system. Even smaller changes have to be extremely well-founded and well-argued, based on thorough evidence and modelling of historic data, which means that the overall framework is destined to be rather more reactive than proactive, and that these reactions will tend to take time. The management of secondary objectives or of other unconsidered aspects can only be improved if it can be shown that primary objectives are not affected. This means that the adaptation of dam management rules (e.g. so that they incorporate drought considerations) is a lengthy procedure.

This said, there is significant capacity, responsibilities and resources to address different issues in a way that does not interfere with the overall framework; there is also the will among actors to address new risks and topics. The district authorities and WVER's approach to the implementation of European directives foresees amicable agreements/cooperation with affected parties, showing high degree of flexibility in on-the-ground implementation. It seems possible to reassign responsibilities in the definition of water resource problems related to flooding, and possibly nature. Resources, however, seem a different issue altogether, with the system quite fixed. The question of available resources seems very important in the final implementation, particularly where municipalities are involved.

Flexibility is also shown in the way that topics pushed by stakeholders have been taken up by the relevant authorities. The question of enabling the return of salmon to the region's rivers was initially pushed by fishermen, who managed to convince authorities to take up these objectives. Regarding implementation and crisis situations, there does not seem to be much flexibility in moving up and down levels, as main decisions are mostly taken by the highest authorities in realising certain plans. Depending on the issue at stake, the decision is often brought up automatically to the superior levels, e.g. the district government.

The ability to include new actors into formal structures of responsibility seems questionable as the structures within WVER and its 'assembly' seem fixed. However, informal relationships are seen to be a way forward in this regard, with new actors being addressed in participation processes, and adjustments to the distributions of responsibility seeming possible.

4.4.4 Intensity

Currently, the relatively weakest point of the governance context for drought resilience policies and measures is its *intensity*. (However, this also holds true in other DROP pilot regions in Northwest Europe, due to the region being overall quite water-abundant.)

The district government seems to constitute the most relevant level in the decision chain concerning water issues related to drought. (It should be remembered that the waterboard only has executing powers, and thus cannot implement on its own accord measures for a new issue such as drought.) At the national and at the *Länder* level, initiatives addressing climate change adaptation have been launched, but are as yet only limited to knowledge exchange and studies. Improving drought resilience has no priority on the political agenda (or not yet at least) and no resources are made available for this topic. According to the district government, the German and North Rhine-Westphalian Adaptation Strategies do not have implications for their daily work, because they are too unspecific to result in concrete requirements and actions. As a consequence, there seems to be a lack of plans or other instruments regarding drought adaptation, as well as a lack of long-term vision for this issue.

This means that the district government is under no pressure due to obligations on the topic of climate change adaptation in general or drought in particular, and nor have the relevant resources been made available. Although they recognise the importance of the issue, they do not see themselves in a position to take it up, and so the district government is currently not driving any process (e.g. establishing its own guidelines, programmes, or implementing adaptation initiatives). It is individual actors that are initiating interesting activities—the DROP project being one of them. The Waterboard Eifel-Rur, a drinking water producer and a hydroelectricity producer interviewed all emphasised the existence of technical projects to enhance the system's robustness, improve risk management and develop backup solutions in case of extreme events. We can say that drought prevention is being addressed in the context of general risk management strategies that use as inputs' precipitation patterns.

WVER can thus be described as the driving force of change in the region. As the responsible for most things water in the Eifel, they are also the first in line to be affected by drought issues, which explains their taking a proactive approach. The overall assessment of the intensity is thus low, but with increasing strength.

4.5   Improving Drought Governance in the Eifel: Conclusions and Recommendations

4.5.1   Conclusions

The observations mentioned in the above section let us conclude that the governance context for drought resilience in the Eifel-Rur can be regarded at the moment as "intermediate". Figure 4.2 shows that the system is overall coherent and shows a fair extent in most governance dimensions, but there is plenty of room for improvements in terms of flexibility and intensity.

This evaluation of the drought governance system as "intermediate" is the result of a general framework which is quite positive for overall water management, but which from a perspective specific to droughts includes interactions which detract from this positive evaluation. We would like to highlight two main ones in this concluding section. First, the system of water rights and the associated water user charges is unduly inflexible, in which it does not allow for creating incentives to reduce water rights. The water rights' system provides strong guarantees for users—in line with a water provision which can offer high security of supply due to its system of reservoirs—but this rigidity has the potential to become problematic both under drought conditions and when faced with longer term impacts of climate

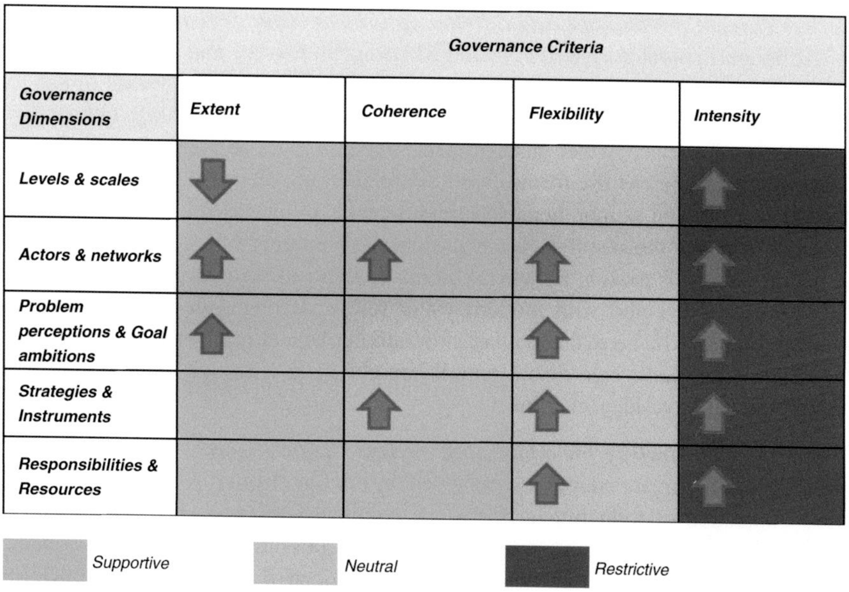

Governance Dimensions	Governance Criteria			
	Extent	Coherence	Flexibility	Intensity
Levels & scales	↓			↑
Actors & networks	↑	↑	↑	↑
Problem perceptions & Goal ambitions	↑		↑	↑
Strategies & Instruments		↑	↑	↑
Responsibilities & Resources			↑	↑

Supportive   Neutral   Restrictive

Fig. 4.2 Summary visualisation—Governance context assessment for droughts in the Eifel-Rur region. *Arrow up* Positive trend in time; *Arrow down* Negative trend in time

change affecting both water availability and water quality. Second, and related to the first point, is the fact that the waterboard's functions and priorities are established by law. Whereas this has positive impacts in a number of areas, this means that droughts—as well as other emerging and novel issues—can only be addressed within the possibilities provided by the current legal framework. Furthermore, the requirements derived from the legal responsibilities mean that changes (e.g. to operational rules) can only be approved after a lengthy review process. This can significantly increase the time lag between issue identification and measure implementation.

These and other governance issues are also addressed in the following section, in which we present possible recommendations to improve the region's water governance in view of droughts.

4.5.2   Recommendations

This section presents possible recommendations to improve, from a drought perspective, the water governance context in the Eifel-Rur region.

1. *Use current possibilities and develop options to manage water demand*
   Although the water system is managed comprehensively and sophisticatedly in the Eifel-Rur, there seems to be a mismatch between the instruments in use for floods, water quality and groundwater,[4] and those addressing quantitative aspects of surface water management (including those relevant for drought purposes). Whereas the former have profited from recent European regulations that have driven comprehensive updates of planning objectives and tools, the latter is rather the result of the historical development of regional water regulations. For this reason, numerous elements seem to some degree incompatible with each other and with modern water resources management. For instance, there seems to be no real incentive structure in place to manage water demand—which would have significant overall benefits from a drought perspective. The options we have identified are:

   (a) *Develop strategy for addressing current inefficiencies*
       From a climate adaptation perspective, but also from a broader governance objective of reducing resource use conflicts and thus enhancing planning security for economic actors, a number of possibilities are currently being missed. These inefficiencies could be reduced if a better use is made of existing instruments that could reduce unused water rights to bring them in line with actual use—including realistic development potential for the local industry in the future. Whereas some instruments to this purpose exist, updating historic water rights in the Eifel-Rur may be resisted by affected users, which means that authorities need to count with political will behind their initiative. They would probably also require an improved resource base to address this extra task over several years, as resources already now seem stretched quite thin.

   (b) *Review water rights and water pricing strategies*
       New, additional instruments which provide adequate steering mechanisms for managing water demand could also be implemented. For instance, current water charges in the Eifel-Rur region are linked to water use, and not to water rights. Including a link to the size of a water right in the charges, for instance by making charges both water rights and water use (e.g. weighting them in an average), could help address current inefficiencies and missed opportunities.
       Interviewees highlighted that owners of water rights would hang on to existing surplus rights for possible future expansion of operations. "Old" rights often provide more legal guarantees than newer ones, which creates unwillingness to trade in old rights for new rights.

---

[4]Strictly speaking, these instruments are not managed by the WVER, as it is not responsible for the relevant groundwater bodies. However, the instruments exist and are implemented by the neighbouring waterboard, which is responsible for the WVERs region's groundwater. For more detailed information cf. Sect. 4.2.2.

(c) *Create incentives to explore alternative water supply options*
Incentives for increased water efficiency (e.g. water recycling) are not felt everywhere, as water recycling comes at a cost (of energy). There seem to be no initiatives in place exploring alternative water supplies, e.g. rainwater harvesting, significant process water recycling, etc. An impulse to increase process water recycling could be given by making creating an economic case (e.g. making it financially beneficial) for the private companies that are the largest water users in the Eifel-Rur region.

2. *Develop a comprehensive and up-to-date database on water rights and water uses*
Related to the previous point are the significant data issues affecting surface water. Up-to-date information would not always be available, both for the different types of surface water rights, as well as for the different types of actual uses of water. Options such as systematic water metering do not seem to be in discussion. The lack of data would be related to the lack of updated legal requirements mentioned in the previous point.
An adequate management of water resources requires comprehensive and up-to-date data on these points. This is a necessary basis for understanding the system and evaluating the potential for increasing system resilience, e.g. by water demand management. Again, a push for data improvement would probably require both political will and to some extent additional resources. The benefits of increasing the water management system's resilience would in all probability far outweigh the expenditures.

3. *Search for synergies between drought preparedness and advisory services/ flood prevention plans*
There seems to be a potential for synergies between measures addressing water scarcity and droughts, and other initiatives being implemented in the Eifel-Rur region. For instance, the possibility of including water quantity aspects in the current advisory services to farmers (within the context of the WFD) would seem promising. Interviewees considered examples such as those of the Somerset region (using moisture sensors to address irrigation needs more precisely and thus reduce water use), in which actors have an economic benefit (reduced costs of irrigation) as very viable.
There is also potential to incorporate drought topics and measures into flood prevention planning.

4. *Authorities' review of decision-making processes: goalposts and stalled processes*
Stakeholders report that some planning processes are somewhat stalled, with little progress over the last 2–3 years. The deadlock would be a result of trying to achieve consensus and keeping to the traditional voluntary approach in topics which are contentious due to significantly different interests and the high price tags of relatively minor concessions. It would seem that the planning process requires a mechanism for addressing these kinds of impasses. Some stakeholders also wished a clearer guidance on the overall process objectives (the

"goalposts") from the responsible authorities. In some cases there may be the requirement for authorities to take a somewhat stronger role. There would also seem to be room for a heightened role for authorities in controlling the on-the-ground implementation of its regulations.

5. ***Develop strategies to maintain in, and add actors to, the planning processes***
   Strategies could be developed to maintain in, and add actors to, the planning processes. Municipalities in particular seem to be finding it very hard to participate in water management, as many are facing extremely significant resource bottlenecks. Particularly, the financing possibilities of any possible measures addressing drought should be given thorough attention. Other actors can be addressed by demonstrating the benefits of particular initiatives, e.g. by local showcasing of the implementation of certain measures.

6. ***Increase synergies with farmers***
   Farmers are a stakeholder group of relevance in the downstream area of Eifel-Rur and that seem to be in a position to impose their own agenda to a significant extent. There seems to be a reluctance to collaborate with water management objectives (e.g. when measures do not coincide with agriculture aims). For instance, municipalities with strong farming presence would resist repurposing some areas of land for WFD Programmes of Measures, although the legal basis is clearly against them. It could be relevant to try to evaluate how to make the relationship with farmers more productive when it comes to drought preparedness so as to avoid this kind of deadlocks in drought planning. An option would be to explore the additional synergies between the waterboard and farmers on water quantity (with a special focus on possible bottlenecks during the summer season), water quality, or on unrelated topics.

Chapter 5
Governing for Drought and Water Scarcity in the Context of Flood Disaster Recovery: The Curious Case of Somerset, United Kingdom

Alison L. Browne, Steve Dury, Cheryl de Boer, Isabelle la Jeunesse and Ulf Stein

5.1 Introduction to Somerset, UK: The Land of the Summer People

Historically, flooding has dominated the physical and political landscape of Somerset, UK. Somerset has been known throughout history as 'the land of the summer people' with the floodplain only being used in the summer, due to its seasonal winter flooding. One of the unique features of this region is the Somerset Levels and Moors—a highly managed river and wetlands system, which is artificially drained and irrigated in order to open the area for productive settlement and uses such as farming. These water management systems extend back to the time of

A.L. Browne (✉)
Geography/Sustainable Consumption Institute, University of Manchester, Rm 1.026 Level 1
Arthur Lewis Building, Manchester M139PL, UK
e-mail: alison.browne@manchester.ac.uk

S. Dury
Community Infrastructure, Somerset County Council, County Hall, PP B2E 2A, Taunton,
Somerset TA14DY, UK
e-mail: sdury@somerset.gov.uk

C. de Boer
Faculty of Geoinformation Science and Earth Observation, ITC, University of Twente,
Drienerweg 99, Enschede 7522ES, Netherlands
e-mail: c.deboer@utwente.nl

I. la Jeunesse
UMR CNRS CITERES, University of Tours, Maison Des Sciences de L'homme, 33 allée
Ferdinand de Lesseps, 30 204 Tours, France
e-mail: isabelle.lajeunesse@univ_tours.fr

U. Stein
Ecologic Institute, Pfalzburger Str. 43-44, 10717 Berlin, Germany
e-mail: ulf.stein@ecologic.eu

© The Author(s) 2016
H. Bressers et al. (eds.), *Governance for Drought Resilience*,
DOI 10.1007/978-3-319-29671-5_5

83

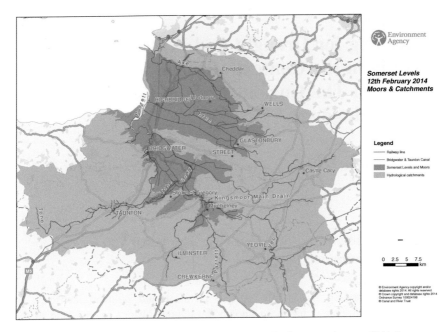

Fig. 5.1 Map of Somerset levels, moors and catchments. Environment Agency (EA) figure

the Norman Conquest (eleventh Century) for the coastal Levels, with the Moors enclosed and drained with fields, ditches, rhynes and engineered rivers between 1750 and 1850 (Clout 2014; Natural England 2013). This landscape has become one of the UK's most significant (peat) wetland natural environments, and has emerged as a result of a complex management history characterised by the coexistence of agriculture and water/environmental management (Natural England 2014). This history has created an interesting heritage of farming, wetland and natural wildlife within the landscape and 13 % of this area is now recognised as a Site of Special Scientific Interest, a Special Protection Area and an internationally recognised Ramsar wetland site (Natural England 2013, 2014) (Fig. 5.1).

Flooding is still a significant agenda for the region with a series of three floods occurring between April 2012 and March 2014—with the flooding event of December 2013 to March 2014 being particularly devastating (McEwen et al. 2014; Natural England 2013; Thorne 2014). However, the area is also sensitive to drought events, having been on the precipice of an increasingly severe drought throughout 2010–2012 as was much of the UK (Lever 2012; Waterwise 2013). Given the theme of this book and the project on which it is based, this chapter focuses on the governance of drought for the Somerset region; however, the assessment and reflections are made with consideration of the broader water management history and current governance structure of the region. In fact the flooding events of 2013–2014 disrupted both the pilot programmes within Somerset presented in Sect. 5.3, and the governance assessment presented in Sect. 5.4. Analytically, this focus on drought in the context of flooding is necessary as recent history has shown Somerset is vulnerable to shifts between extreme events even within one year (e.g. in 2012 there was a shift from drought to small flood events). This chapter reflects on the governance conditions for drought following two visits to the region—one in September 2013 following the period of drought, and another again in October 2014 after a period of flooding recovery (Browne et al. 2015).

The chapter is structured as follows: Sect. 5.2 overviews the national water management and drought context; Sect. 5.3 highlights the geo-hydro context and overviews specific drought policy and measures taken in the Somerset region with the non-academic partners in the DROP project. Section 5.4 in particular captures the assessment of the governance context in Somerset. It highlights our reflections after the first visit that there were many positive elements emerging. These ranged from an increasing breadth and variety of instruments and measures used to plan for drought; increasing number of relationships being built to deal with policies and on-the-ground measures for drought; and increasing visibility of the issue of drought for the region after a period of extended dryness, and as a result of awareness raising activities of a number of stakeholders. The implication of governing for drought in the context of flooding recovery is also discussed related to the flooding period of 2013–2014 discussed in Sects. 5.2 and 5.2.4. In Sect. 5.5 we conclude by exploring the potential meta-governance failures in the wider English water management system. In particular, we highlight the development of political 'silos' and fragmentation that were expressed in situ in Somerset in the aftermath of the flooding events. Policy and implementation silos exist between drought and flood in the definition of the target of adaptation efforts for a future of climate change. These silos need to be addressed in ongoing water management policy, and on the ground adaptation actions in the Somerset region if resiliency to future events is to be increased.

5.2 National and Regional Climate Change, Water Management and Drought Governance Contexts

5.2.1 The Geo-Hydro Context of Somerset Water System and Future Climate Impacts

The Somerset Levels and Moors is a unique manmade wetland landscape of international importance for nature and archaeology. A significant part of the low-lying Somerset Moors is designated as a Special Protection Area and a Ramsar site, which depend upon flooding. The area is also rich in archaeological sites that depend on waterlogged conditions for their preservation.

The steepness of the uplands, coupled with the geology and soil conditions, generates quick run-off from short intense rainfall. The upland areas of the wider catchment (Mendip, Blackdown and Quantock hills) are very steep, but the lowland areas of the Somerset Levels and Moors are very flat. This means that rainfall run-off travels very quickly down from the uplands but then slows down and pools in the Somerset Levels and Moors. The high-level embanked channels overflow and floodwater is stored in the Moors before it can reach the Estuary. In addition, the very shallow gradient on the Somerset Levels and Moors means that the area drains water away very slowly and relies on a complex network of pumped drainage channels. Tide locking is a particular feature of the Somerset Levels and Moors; the lower reaches of the rivers Tone and Parrett are tidal for some 30 km (18.6 miles) from the Severn Estuary. The capacity of these channels can be significantly reduced by high tidal conditions; in particular the Parrett as it has no tidal sluice or control structure.

Widespread flooding of the lowland moors happens regularly from the perched main rivers which run through them. The moors are protected by raised defences as are many of the small villages and communities. During the 2013/2014 fluvial floods, flood defences across the area protected over 200 km^2 of land and over 3500 properties. However, large areas of land were still flooded for many weeks and these included 172 properties. Strategic infrastructure which included main roads and the rail network were affected badly and some small communities were cut off for many weeks.

Climate change increases both the risk of flooding and drought in Somerset. UK droughts are projected to be more severe and affect larger areas of the country over the next 100 years. Example studies include a publication in the journal Water Resources Management (Rahiz and New 2013) and a study by scientists from the European Commission's Joint Research Centre and Kessel University (2014). Climate change will also increase the potential for stronger rainfall events. The implications for flood risk, however, will vary widely from location to location depending on local climatic changes that are at present difficult to predict with confidence. Climate change may also result in changes to large-scale atmospheric circulation patterns like jet streams, which are harder for climate simulations to predict. Recent results with state-of-the-art climate models have raised the possibility that climate change may affect the jet stream more than scientists previously expected, making floods in the UK more likely. However, the uncertainty in these projections remains large. Flash flooding could also become more frequent as extreme rainfall events are consistently predicted to become more severe.

5.2.2   Regulatory and Governance Context of English Water Management

In 2008 the UK government ratified the Climate Change Act. One aspect of this piece of legislation ensures relevant public bodies put plans in place to adapt to climate change. This has involved a range of activities specifically focused on adapting to climate change such as the National Adaptation Programme, and the UK Climate Change Risk Assessment (e.g. HM Government 2013; Wade et al. 2013). For example, the UK government now has 'Adaptation Reporting Power' requiring a range of stakeholders and companies to provide detailed reports on the current and future predicted impacts of climate change on their organisations, and their proposals for adapting to climate change. Many of these programmes and policies have influenced climate change adaptation planning and implementation within water management settings in the UK and in Somerset. There have been a wide range of research papers and policy white papers in the water sector assessing the climate change impacts on the UK water sector (e.g. Defra 2013; Fenn and Wilby 2011; HM Government 2008, 2011, 2012, 201; Wade et al. 2013). Relatedly, the water industry in the UK has engaged with a range of activities related to climate change forecasting and adaptation plans as part of their water resource management plans (WRMPs are also part of the Water Framework Directive—WFD—reporting requirements).

Water management in the UK reflects a complex, multilayered and multi-actor regulatory and governance system. Water resources are managed differently across country boundaries in the UK; therefore, the following description applies to England only. Defra (Department of Environment Food and Rural Affairs) has the overall responsibility for policy related to the water and sanitation sector (e.g. quality of drinking water and other waters, sewerage treatment and reservoir safety). Defra also ensures that the "legislative framework for water management is fit-for-purpose" (Environment Agency 2015a, p. 7). The executive body who has responsibility for environmental regulation (long-term planning for quality, water provision, climate change adaptation, WFD implementation) is the EA. Economic regulation of the water industry falls under OFWAT (Office of Water Services), and monitoring of drinking water quality is covered by the Drinking Water Inspectorate. Natural England (NE) is an executive non-departmental public body responsible to the Secretary of State for Environment Food and Rural Affairs with a remit to manage and adapt areas of natural significance, and to manage green farming schemes in England.

In Somerset, Irrigation Drainage Boards (IDBs) oversee district water level management (of the Levels and Moors), and they also work on water level management in order to reduce flood risk to property and people. The IDBs are co-defined as Risk Management Authorities within the Flood Water and Management Act of 2010 alongside the EA, local authorities and water companies (Association of Drainage Authorities, no date). Defra is responsible for the IDBs; however, they work closely with the EA and lead local flood authorities, and are funded by a range of beneficiaries only one of which is the government. The Somerset Drainage Board Consortium (SDBC) is responsible for managing water levels to protect people, the environment and property (SDBC 2014). The English drinking water and wastewater companies are privatised but are still regulated by, and have reporting responsibilities to, the governmental bodies identified above—Defra, EA, OFWAT and the DWI (Water UK 2015).

Somerset County Council (SCC) is responsible for managing strategic local services in Somerset. With regards to water management, they have a role to play in emergency planning, consumer protection, town and country planning and local flood management. They also act as 'knowledge brokers' for climate change adaptation awareness raising and other activities. Finally, there are a range of NGO's (non-governmental organisations) and associated stakeholders who have fairly large stakes in taking forward adaptation planning for climate change and water management within the Somerset region. A lot of these groups' interests are reflective of the delicate balance between nature and agriculture in the region and include, for example, the Royal Society for the Protection of Birds (RSPB), Farming and Wildlife Environment Group, South West (FWAG, SW) and Somerset Wildlife Trust (SWT). In the context of this current project, FWAG SW, RSPB and SWT are sub-partners to the Somerset regional pilot.

5.2.3   Drought Governance Context: Managing Water During Normal and Crisis Periods

Given the complex physical and policy landscape for water management in England generally, and in the Somerset area more specifically, it is useful to reflect upon drought governance as it relates to management of drinking water, agricultural water, and water for nature and biodiversity. Drought is seen to both influence, and be influenced by, activities related to drinking water, agriculture and horticulture, industrial activity, infrastructure (e.g. energy provision), navigation, and environmental protection (fisheries, wetlands, wildlife and plants) (Environment Agency 2015a).

There are a range of planning activities to increase the resilience of the English water system to short and longer term changes as a result of climate change and increasing water demand, including actions specifically related to exceptional drought events. For example, water companies in the UK are required to consider both supply planning (such as new supply and transfer investments) and demand management (water efficiency activities, and temporary use bans to restrict household and business consumption during droughts) as a way to increase the resilience of the water system to social, technological and climatic change. Achieving a resilient supply and demand balance has been deemed necessary as drought and water scarcity (van Loon and van Lanen 2013) are historical features of the water system in the UK (Rahiz and New 2012; Taylor et al. 2009) and because climate change will increase the frequency of short-term periods of dryness and multi-year droughts (Environment Agency 2015a).

The other actions that ensure the UK water system is resilient to changes in climate are how water is managed during drought and water scarcity events. With regards to drought governance at the national level, there is an existing emergency management hierarchy of national, regional and local decision making around emerging periods of crisis. The drought management scheme that covers Somerset is the 'South West Drought Plan' (Environment Agency 2012a), supported by a national level drought framework, which was updated in June of 2015 (Environment Agency 2012b, 2015a). The 2010–2012 drought also led to a recognition of the need to strengthen the networks that need to be mobilised during drought events (e.g. Lever 2012; Waterwise 2013). In 2012 a continually functioning national drought group was set up involving stakeholders from NE, NFU (National Farmers Union),

Water UK, water companies, RSPB and many other organisations to address this issue.

The development of a new national level drought plan in 2015 is significant for the governance assessment, which will be presented in Sect. 5.4. For example, when we visited the region in 2013 a number of stakeholders indicated that there was a lack of clarity in the way problems were defined and goals related to drought were set, as well as lack of clarity over how particular drought strategies and processes are implemented, and by whom. This created a sense of fragmentation of roles and responsibilities of regional responses to drought within the Somerset region.

The EA report "Drought response: our framework for England" (Environment Agency 2015a) provides a better outline of the responsibilities for managing water resources during periods of drought. The main organisations responsible during a drought include

- "the EA; provides strategic oversight and responsible for monitoring, reporting, advising and acting to reduce the impact of a drought on the environment and water users
- water companies; responsible for managing water supply for their customers and taking a range of measures to maintain supplies whilst minimising environmental impact
- government; responsible for policies relating to water resources.
  A number of other organisations and groups also play an important part in managing drought, including NE, Canal and River Trust, local councils and representative bodies such as National Farmers' Union (NFU), UK Irrigation Association and environmental charities. All those involved in dealing with the effects of drought plan their activities in case a drought occurs and ensure that the responsibilities of different parties are clearly defined and understood". (Environment Agency 2015a, p. 6).

During a drought the EA carries out a range of actions at a variety of scales depending on the nature of the drought (Environment Agency 2015a). Drought incident teams decide on courses of action including a range of environmental, hydrological and social-economic indicators to categorise drought; assess short- and long-term forecasts of drought; convene strategic drought management groups relevant to the scale of drought (e.g. National Drought Group); act as rapporteur to a range of governmental actors, partners, water companies, stakeholders and the media; deal with drought orders and permit applications from water companies; implement environmental restrictions on abstraction licences; and provide clear advice to the government (Environment Agency 2015a).

With regards to other actors during droughts, the role of Defra is to work with the EA and water companies to maintain public water supplies and minimise damage to the environment (Environment Agency 2015a). Water companies should be prepared for periods of dry weather through a range of plans called

Drought Plans. These Drought Plans include a range of actions during droughts including publicity campaigns, customer restrictions, drought permits and orders changing normal operations and abstractions. They must match the long-term strategic adaptation strategies outlined in their Water Resource Management Plans, and satisfy 'needs' of both the environment and various water users (Environment Agency 2015a). The role of local councils—as discussed above— is more in regards to emergency management and contingency planning, and convening local resilience forums (Environment Agency 2015a). The role of NE during a drought period is to give advice regarding the influence of the drought on a range of 'nature indicators' including protected habitats and species. NE also has a role to play in providing advice to a range of stakeholders (industry, farming, local community, etc.) during a drought period (Environment Agency 2015a). As NE also manages a number of National Nature Reserves they also have responsibility for a range of actions to protect vulnerable species in these areas (e.g. drought monitoring, restricting access to vulnerable nature areas) (Environment Agency 2015a).

Fig. 5.2 Flooding in Somerset county in february 2014. We can see the St Michael Church on the hill isolated from the flooding territory (Somerset County Council)

5.2.4   Flood Policy Developments in Somerset Since the Floods of 2013/2014

Although this book and project are focused on drought governance and resilience, the flooding events that occurred in Somerset across 2012–2014 have significantly influenced the long-term water management futures of the Somerset region. Thus, it is also useful to overview the developments to flood policy and how they have influenced adaptation actions towards drought and water scarcity governance.

The flooding in 2013–2014 included the flooding of the Parrett and the Tone catchments. It was the largest flood event known to have occurred in the region in the last 250 years with the army being deployed to assist during the crisis (BBC 2014; Environment Agency 2015b; see also Fig. 5.2). The floods became a hot political topic during the immediate crisis and recovery, with national and European media presence covering the contested and strongly debated causes, and solutions, to the problem. Thorne (2014) reflected how the 2013/2014 floods were socially divisive, with our own research during the period of flooding recovery highlighting the de-legitimisation of various stakeholders involved in water management throughout these public debates. It can be argued that the flood event obscured some options for recovery and future adaptation and entrenched others (Browne 2014; Butler and Walker-Springett 2015). The flood event also offered an opportunity for improving resilience, resistance and relations between the public, various regional stakeholders and other water practitioners (Butler and Walker-Springett 2015). The policy developments that emerged as a result of these flood events need to be viewed in the context of the divisive and contested nature of the flood recovery process, throughout which there was a series of social and political struggles to maintain a more or less balanced policy response.

The EA, whose funding and investment decisions are based on national assessment and cost benefit analysis, is currently the main provider of flood risk management activities. After the flood event of 2013/14, the 20 Year Flood Action Plan was developed to achieve a long-term vision for the area to reduce the extent and impact of flooding (Cameron 2014). The plan includes works being done to repair flood defences in the region, build new flood defences and enhance the capacity of various drains in the region to reduce the risk, depth and duration of any floods in the future (Environment Agency 2015b). It also identified dredging around the Levels and Moors to reduce flood risks (Environment Agency 2014), creating new banks, increasing pumping capacity and additional maintenance activities such as weed control (Environment Agency 2015b). The implementation of the 20 Year Flood Action Plan is being spearheaded by FWAG SW (a sub-partner on the DROP project).

The Somerset Rivers Authority (SRA) is a key part of the 20 Year Flood Action Plan. The SRA will bring together the Flood Risk Management Authorities (the EA, the Internal Drainage Boards, the Lead Local Flood Authority (SCC) and the other Somerset local authorities), to provide a strategic overview of the continued delivery of the Flood Action Plan, and to develop, agree and publish a Common

Works Programme. It was set up to deliver greater local control and responsibility for maintaining and improving water and flood risk management on the Levels and Moors and across Somerset. The SRA will be a new body with its own Board, which will include representatives from each of the following partners: the five District Councils, SCC, the EA, the Parrett/Tone and Axe/Brue IDBs, the Wessex Regional Flood and Coastal Committee, and NE. Interim funding of £2.7 m has been secured for the SRA for the 2015/16 financial year, the majority coming from Defra and DCLG (Department of Communities and Local Government). Local partners in Somerset, Defra and DCLG are working together to review options for a sustainable local funding solution for the work of the SRA from 2016/17 onwards.

5.3 Drought Measures Taken Within Somerset in the Context of Flooding Recovery

Concurrently to the 2013–14 floods, and the emergence of new policies and actions related to flood risk management, the regional partners on the DROP project (SCC, FWAG SW, SWT, RSPB) were developing a range of on-the-ground measures to enhance local resiliency to drought and water scarcity in the region. This section overviews the specific drought adaptation actions for agriculture and nature taken in the DROP project by the regional partners SCC and subcontractors FWAG SW, and SWT across 2013–2015.

5.3.1 Agriculture and Drought Resilience

The changing rainfall patterns under a changing climate are likely to have a profound effect on land management and farming in Somerset. This will result in additional demand for winter storage of water both to alleviate flooding and also to cope with reduced summer rainfall. Many farmers may need to consider irrigation from this winter storage, which could be anything from floodplain retention areas and creation of wetland habitats, to interception ponds, collection pits and butts. Farmers will need to implement water conservation measures and explore innovative approaches to water management on farms. In DROP, FWAG SW has helped farmers and landowners to adapt to these increasing extreme rainfall patterns.

As a county, Somerset has the greatest variety of soils in England. A soil risk assessment across the whole county to identify soils at risk of drought is missing and would be valuable in determining the actions required to make rural areas of the county more resilient to drought. FWAG SW has done a lot of work on Soil Risk Assessment for flooding and run-off. Within the DROP project they have altered this approach and criteria to do the same for drought. This provides valuable information and targeting for other measures identified within the project.

As previously discussed, in response to the 2014 Somerset Floods, FWAG SW has been involved in putting together a 20 Year Action Plan for water management in Somerset. This has included many of the Natural Flood Management and Soil Management measures that are included in the DROP Project. The Land Management aspects of the 20 Year Plan include a range of interventions like woodland planting, increasing soil organic matter, run-off attenuation features, improving soil structure and slowing watercourse flow that will improve drought resilience. This has constituted a major part of the progress achieved by the DROP Project in 2014.

Four sub-projects undertaken by FWAG SW as a sub-partner on the DROP project are related to improving the storage, conservation and recycling of water on farms; improving soil organic matter and soil structure; developing modelling and technology transfer for irrigation scheduling and water application management; and developing an Area Level Water Management scheme. The four projects are as follows:

(1) **Working with farmers to investigate ways of storing, conserving and recycling water for on-farm use**. This has involved implementing water efficiency measures and water conservation techniques in land-based businesses. Four 'demonstration' farms showcase various in situ soil protection measures, and open days have been held to encourage other farmers to implement these measures on their own land. Measures include (i) the reinstatement of ditches and drains to slow and elevate run-off on a historic rural estate in Somerset; (ii) use of temporary grassland to minimise soil erosion and installation of filter fences to prevent soil from washing into neighbouring properties; (iii) installation of a stone gabion and fencing feature to hold back fine soil that washes from the gently sloping arable field; and (iv) soil bunds to prevent soil washing through a hedge and a newly installed filter fence, silt pond, drains and established winter sown oats after maize at a local farm.

(2) **Investigating ways of improving soil organic matter levels and soil structure**. Different types of cover crop have been trialed on two pilot sites to help build organic matter. Healthy soil structure and high organic matter levels help to increase soil resilience against the effects of waterlogging and drying. This is especially important on arable farms where the normal sources of organic matter are in shorter supply due to the removal of biomass. Both sites are being monitored through a combination of infiltration measuring and earth worm counts.

(3) **Developing modelling and technology transfer in the Upper Parrett catchment on irrigation scheduling and water application management**. The Upper Parrett was chosen as a pilot area because water demands for agriculture here are high related to potato production. This study aimed to identify opportunities to improve the accuracy of irrigation scheduling to deliver potential savings during summer months within the Upper Parrett catchment when available resources are at greatest risk. In conjunction with

potato producers Branston Ltd, and following installation of Dacom probes in the four fields in May 2014, Soil Moisture Deficits (SMD) have been monitored throughout the growing season to determine relative dryness within the soil profile and enable the growers to more accurately manage irrigation water applications. Early indications are that the Dacom probes have provided a more reliable guide of actual SMDs and enabled growers to make more informed decisions on actual SMDs rather than relying upon guideline figures and subjective visual assessment of soil dryness. A calculation of the financial value of the harvested crops will be completed in order to provide baseline data to assess the cost effectiveness of the different SMD monitoring techniques employed.

(4) **An Area Water Level Management scheme for the East and West Waste area of the Somerset Levels and Moors developed in 2014**. It assesses the current standards of water level management and watercourse conditions, and considers existing and future pressures from drought, as well as development and agriculture. A local contractor was employed (via the local Drainage Board) to carry out water level management works in the study area to reduce water loss and improve drought resilience.

Fig. 5.3 RSPB site preparation for new sluice installation, West Sedgemoor, Somerset UK

5.3.2 Nature and Drought Resilience

Other activities undertaken in Somerset during the DROP project have focused on land management advisory work that is connected to drought-proofing vulnerable areas, and capital investment in infrastructure to drought-proof key vulnerable areas on the peat moors and on the clay levels. This increases the ability to cope with drought conditions. Low-lying inland areas of Somerset depend on water management to maintain their environmental features and agricultural interests. Areas with exposed peat soils are particularly vulnerable to drought, which can damage peat soils, affect agricultural production and impact the natural and historic environment. Innovative landscape-scale approaches to water management will be used to plan and implement changes in water management for these areas.

As part of DROP, RSPB (through its subcontractor, SWT) has worked on nature reserves to make the most vulnerable habitats more drought resilient, restored habitats on nature reserves that contribute to drought resilience in the landscape and engaged with private landowners to pilot drought resilient restorations of peat extraction sites. Various water control structures were installed on West Sedgemoor RSPB reserves and SSSI between July and December 2014, including 9 culverts, 4 penstock flapvalves, 2 tilting weirs and the removal of 12 old structures (Fig. 5.3). The installation of these new structures is part of a programme of work

Fig. 5.4 Restoration of reed bed in Westhay Moor, Somerset, UK

which improves efficiency and functionality of water management on this large wet grassland site to better cope with periods of both water shortage and water surplus, of which the DROP work is a significant part.

Until the Middle Ages large parts of the Brue Valley were covered in a peat-forming raised bog. In the UK the area of lowland raised bog is estimated to have diminished by around 94 % largely through agricultural intensification, afforestation and commercial peat extraction. Future decline is most likely to be the result of the gradual desiccation of bogs damaged by a range of drainage activities and/or a general lowering of groundwater tables. In the Brue Valley the majority of the raised bog has been lost to peat extraction and agricultural intensification. The remaining fragments, which are now all within nature reserves, are raised above the surrounding peat voids and are consequently very difficult to keep wet. The largest remaining fragment of raised bog belongs to SWT, who have undertaken work to improve the habitat's resilience. Through the DROP project SWT has extended the programme of tree removal and scrub clearance; improved structures, fencing and gates to improve the grazing regime; and improved the ditch network and bunding to extend the areas SWT can deliver water to. A significant amount of restoration work has been carried out on the Westhay Moor raised mire habitat, including the clearance of 2 ha of scrub. A reed bed on this site was also restored by cutting channels and installing a structure to allow better water circulation (see Fig. 5.4). These actions will increase the drought resilience of both these habitats.

5.4 Assessment of Drought Governance in Somerset

Following the first visit to Somerset by the authors in 2013, we noticed several differences compared to many of the other case study regions presented in this book. There was much more awareness of climate change and its potential dual effects on water levels, with a wide range of stakeholders engaged in adaptation projects across the region. Our major reflections after this visit was that there was some fragmentation in how roles and responsibilities were defined, particularly related to initiating engagement and actions during drought periods. As discussed in Sect. 5.2 this may partially be resolved through the clearer responsibilities outlined in the new 2015 national drought response framework (Environment Agency 2015a).

However, the 'seismic shock' of the 2013/14 flooding altered the status quo for the discussions on water management. The politicisation of flooding in the region led to a reinterpretation of water management that became far more one-sided, with initial policy and practical measures focused largely on engineering type approaches such as dredging. Multiple stakeholders quickly called for the creation of more discharge capacity, a call that was magnified multiple times by the media and politicians. Although the Flood Action Plan was eventually developed—and the partners involved have pushed a more integrated, catchment management approach that will also increase resilience for drought and water scarcity—the lack of committed funds for fundamental aspects of this plan (such as to support the activities of the SRA) is problematic.

This section reflects briefly on the details of the governance assessment made as part of the DROP project by the authors, and reflects upon the four qualities of governance (extent, coherence, flexibility and intensity) underpinning drought adaptation within the region. As discussed in Chap. 3, (1) extent means that all elements are taken into account, (2) coherence means that elements of governance are more reinforcing and not contradictory of each other, (3) flexibility is that multiple roads to achieving goals are permitted and supported and (4) intensity occurs when the governance context urges changes and improvements in the status quo or current developments.

5.4.1 Extent

First, a high level of extent was observed in terms of stakeholders since their involvement in the management of drought in Somerset region was very strong and positive (cp. Sect. 5.2). The governance assessment revealed that the relationships that do exist around drought are largely positive and these relationships were also seen as having improved as a function of experience in the 2012 drought. The extent and nature of these stakeholder networks have been further clarified in the national drought framework released by the EA in June 2015 (Environment Agency 2015a). While the flood of 2013/14 was problematic and devastating for many, it can also be argued that it positively enhanced the range and scales of actors and networks involved in the issue of water management for the region, and the strategies and instruments that were being adopted to deal with water management particularly at the catchment level.

We found that there was a proactive anticipatory approach to drought management across water supply, nature and agriculture sectors in the Somerset region. This is reflected in the sorts of activities brokered by SCC and the sub-partners of the DROP project described in Sect. 5.3, and in the way that the partners involved in DROP have also been key actors in the 20 Year Flood Action Plan and SRA. Our assessment found that the types of strategies being suggested for increasing drought resilience were comprehensive for drinking water, but that a larger range of strategies for agriculture and nature specifically related to drought and water scarcity adaptation still need to be developed and implemented.

In terms of the future, a complex and changing fiscal context for spending on environmental issues such as climate change adaptation, water management and flooding (Committee on Climate Change Adaptation 2014), including ongoing reduction in overall funding of local authorities, EA and DEFRA may potentially restrict the adaptation activities possible within the system. For example, the 20 Year Flood Action Plan and the SRA (and other such related partnership activities for water management occurring in the region) will rely on financial partnership investment in order to access government funds for ongoing activities (Committee on Climate Change Adaptation 2014).

Despite this, the Somerset Water Management partnership did manage to get funding for a catchment management approach, via a successful funding bid from DEFRA's 'Catchment Partnership Fund' for the financial year 2015/16. This fund supports eligible organisations seeking to host a new or existing catchment partnership. FWAG SW was awarded £11 K as a catchment partnership 'host'. The money is primarily to demonstrate to Defra that there is movement in the direction of a stakeholder-led catchment plan. This funding is not for delivery; it is to enable catchment partnerships to be formed (with the wide range of stakeholders interested in water management). A Working Group has been constituted to plan how to work with communities and stakeholders to investigate mutually beneficial solutions to problems faced by the catchment. The Somerset Water Management Partnership exists to promote a sustainable and integrated approach to water and land use management in Somerset's catchments wherever possible. It provides an over-arching, broad-based advisory and consultative forum in which all aspects of water management in Somerset's catchments can be discussed and consulted upon.

5.4.2   Coherence

Second, there was coherence in the different levels and scales of stakeholders involved in drought governance in the region, and a positive coherence across different actors. These include effective statutory relationships but other examples also exist of partnerships going beyond the regulatory remits to work together. This collaborative way of dealing with drought and water scarcity is increasing in importance. These relationships became clearer after the drought period of 2012. The coherence of the stakeholders involved in drought management (particularly during periods of crisis) was also further clarified in the 2015 national drought framework (Environment Agency 2015a). Furthermore, although the floods can be seen to have been socially and politically divisive, constructive activities such as the catchment management approaches suggested in the Flood Action Plan and the sorts of relationships being initiated at a catchment level to respond to these plans could also be leveraged to support future drought adaptation activities.

As a result of previous experiences of drought and water scarcity there was an acceptance that drought was a problem for the region. There were also fairly consistent definitions of the problem and the goals related to drought and water scarcity across the different stakeholders, including farmers, although this idea of drought as a problem became increasingly restrictive following the floods particularly for non-specialist, urban and regional publics. However, key actors that we interviewed following the floods still identified drought as a potential future problem for the region (although potentially secondary to flooding as a problem).

There is a potential incoherence between the strategies suggested for flooding, and those for water scarcity. There also seems to be a question about coherence in regards to the measures, strategies and instruments to deal with drought and water scarcity in the region. The sorts of approaches being suggested for flood recovery and resilience (e.g. in the Flood Action Plan) for example, could potentially include strategies that are also positive for drought resilience, yet the extent to which these perspectives are combined is unknown. The responsibilities and resources for coherent drought and water scarcity governance were also seen to be more restrictive and decreasing as a result of the attention being pulled towards the 'primary' issue of flooding in the region. Stakeholders' responsibilities and resources were being pulled towards this issue, rather than there being explicit policy development to support greater coherence between flooding and drought policy and action.

5.4.3 Flexibility

Third, in regards to flexibility a positive assessment was made of different actors and networks that were involved in drought adaptation in various ways. However, the flood experiences were observed to have eroded some of the legitimacy of the actors in the region (such as the EA). This assessment therefore moved from a positive sense of flexibility related to multiple actors involved to achieve drought resilience and adaptation, to one that was slightly more restrictive. Despite this there is still a fairly clear institutionally defined approach to the problem of drought management, and there is some flexibility within the goals (see Environment Agency 2015a). Some actors, however, suggested that there was a lack of clarity in the way problems were defined and goals were set, as well as how they are implemented into particular strategies and instruments for drought adaptation (discussed in part in Sect. 5.2).

For example, the definition of the problem of climate change adaptation as flood recovery and mitigation diminished the legitimacy with which lead actors could talk to other actors about drought and water scarcity management (e.g. the county council talking to farmers or citizens, the EA engaging stakeholders in continuous talk about drought in the context of flooding recovery). Therefore, framing climate change adaptation as recovery and mitigation from flooding reduced the flexibility with which actors could take the lead in pushing forward a water scarcity and drought adaptation agenda in the region in the period of flood recovery. The governance assessment made in 2013, however—reflecting on the period of drought experience in 2012—did reveal that the lead strongly sits with a range of actors when dealing with drought management policy and processes (if there is an actual drought). The process of trigger points and responsibilities is clearly defined. However, there was one criticism that such a defined approach for what happens during a drought can lead to an inflexibility at a local level of the plans. The recent drought in 2012 provides some interesting examples of the lead shifting between actors for pragmatic and strategic reasons, particularly related to the communication of drought. This has been further clarified in the 2015 national drought framework.

Table 5.1 Final assessment of governance context for drought in Somerset, UK after two field visits (and after Winter 2013/2014 floods)

	Criteria			
Dimensions	Extent	Coherence	Flexibility	Intensity
Levels and Scales	Positive and Increasing	Positive and Increasing	Neutral and static (was supportive but has decreased post floods)	Neutral and static (was supportive but has decreased post floods)
Actors and Networks	Positive and increasing (increasing with other forms of water mgt)	Positive and increasing (increasing with actors going beyond duties)	Neutral and static (was increasing but is now on hold post floods)	Neutral and static (was increasing but is now on hold post floods)
Problem perspectives and goal ambitions	Restrictive and static (was supportive but has now decreased specifically for drought mgt after floods)	Restrictive and decreasing (was supportive but has now decreased specifically for drought mgt after floods)	Restrictive and decreasing (was neutral assessment but is now less flexible after floods)	Restrictive and decreasing (was increasing in intensity but has now decreased specifically for drought mgt after floods)
Strategies and instruments	Positive and increasing	Neutral (many instruments but implementation lacks coherence)	Positive and increasing	Positive but static
Responsibilities and resources	Positive but decreasing (after floods)	Restrictive and decreasing (after floods)	Positive but static	Negative but static
Key for Colours. Red: Restrictive Orange: Neutral Green: Supportive				

5.4.4 Intensity

Finally, there is a strong sense of intensity for drought issues from all levels and scales in the region. The lead actors and networks consider drought management and water scarcity as part of their core business. This is both as a result of a regulatory environment that requires this (see Sect. 5.2), and an example of how actors are going beyond regulatory defined agendas to promote the issues across a range of stakeholders. There is a strong intensity seen in the use of instruments and measures, and a process of constant renewal of the plans for drought in the region. Where the intensity of the issue decreases is in the sharing of resources associated with the tasks of adaptation. It is likely that this disparity between the intensity of problem definition and actual resources will only increase in the face of the reduction of public funding under conditions of austerity of the current government. These issues of problem definition for drought need to be seen in a complementary way with that of flooding which is explored below in some more detail.

On the second visit after the floods these assessments were mostly 'flattened'. By this it is meant that the intensity with which certain levels or actors were pushing for change and management reform specifically for the issue of drought and water scarcity had been greatly decreased (and defined as neutral and static in Table 5.1). Where it became particularly problematic was in the problem definition and goal ambitions as a result of the experiences of the 2013/2014 drought. The catchment management approaches being suggested as a form of flood mitigation and adaptation may potentially include drought measures in terms of the definition of drought as a problem for the region. Even the ability of different actors and stakeholders to define drought as a problem for the region became very awkward politically and socially as a result of the floods.

Table (5.1) provides a summary of the final assessment of the governance context for drought and water scarcity as a result of the two visits, and reflects the change in direction for many of these criteria as a result of the 2013/2014 floods.

5.5 Conclusions: Planning for Adaptation in the Context of Contested Material Water Histories and Meta-Governance Failures Within the Broader Water Sector

The case study of Somerset shows how the material water histories of a region—and the ways in which these histories are governed through both emergency management and longer term planning processes—shape the directions of flood and drought adaptation. The Somerset case study is unique as it offers an opportunity to explore in situ the political contestation that can occur around water management and climate change adaptation, in particular, the siloing of policy areas, and the fragmentation of adaptation activities this can create.

In part this has to do with the way that the flooding events of 2013/14 played out in time. The Somerset flood event reflected a departure from the normal conceptualization of flooding as an extreme event, with a 'blame game' about the causes of the flooding and the extent of the impacts being initiated against a number of key governmental stakeholders (McEwen et al. 2014). Emergencies and extreme events reflect decisions that are both planned for (contingencies such as the National Drought Framework discussed in Sect. 5.3, or the UK Civil Contingencies register), and sets of decisions orchestrated at the time of an emergency or exceptional event (Adey and Anderson 2011). The form that the 'flood intervention' took during this period was highly politicised (McEwen et al. 2014), and reflected a militarization of emergency water management. This is partly to do with the unexpected scale and duration of the flood event. Also, as Gilbert (2012) states, such militarization of climate change events "perpetuates an externalised concept of nature that is to be commanded and controlled, with no real sense of ecological prioritization" (p. 10). The discussions following the floods were dominated by conversations of infrastructure and engineering solutions such as dredging as longer term adaptation options (Browne 2014). Catchment management for example—although promoted by WFD, Defra and in the Flood Action Plan for the region (Cameron 2014)—did not strongly feature in the development of formal policy in the region following the floods (Environment Agency 2015b). Lead stakeholders in the Flood Action Plan and SRA are, however, now promoting this agenda.

Recent research has shown that the winter floods increased British peoples' perspectives that climate change was happening in the UK (Capstick et al. 2015). It is a missed opportunity that such events were not used to encourage a public discussion about the potential extreme climate and water futures of both flooding *and* drought for the UK in public fora. A consequence of the policy outfall of this extreme event was a siloing of flooding from wider water management issues, partnerships, networks and adaptation activities already occurring in the region and which had harnessed a more complete sense of the connection of drought and flooding governance than that represented in the policies that emerged post the floods of 2013/14. Encouragingly, this recognition of the need to integrate flood and drought research, innovation, practice and policy development is now emerging in the industry (see for example Wharfe (2015) UK Water Partnership report).

What is important to reflect upon, however, is the way that the flood recovery and policy developments were framed, and how dealing with the flooding as an extreme event in this way closed down certain lines of discussion about how to connect water management adaptation activities in a more comprehensive way. Butler and Walker-Springett (2015) have reflected on the complexities of the ways floods are discussed within media and public fora, and how this obscures lived, private experiences which then shapes particular (restricted) policy and implementation activities. This is also reflective of other work on the social dimensions of flooding recovery (Medd and Marvin 2008; Whittle et al. 2011). Recovery from extreme events is often framed in terms of infrastructural recovery and adaptation; however, these are also missed opportunities to connect to a fuller understanding of infrastructural resilience (e.g. connecting across areas of water management and promoting climate change resilience more generally), and in creating greater community resilience (e.g. as discussed by Medd and Marvin 2008, 2014; Whittle et al. 2011, 2012).

Engagement with the critical literatures on water management in England highlights that the siloing of these water management areas in this way (especially during crisis) is not an unanticipated consequence of the existing water governance system. Both scarcity and flooding events in the UK continue to be framed as mismanagement and governance failures. This extends beyond the case study of Somerset and reflects the nature of water management in England. For example, drought events in England have historically been understood as a failure of planning and as such are seen as reflective of industry incompetence and a lack of planning (e.g. Hope 2012). The need for greater investment in supply and sewerage systems in order to create greater short and long-term resilience in the water industry was in fact a large part of the justification for the privatisation of the water industry in England (Bakker 2003, 2005; Maloney and Richardson 1994; Medd and Marvin 2008; Moss et al. 2008). Walker (2014) argues that the forms of water governance that have emerged from these conditions have actually created a (meta) governance failure when it comes to the governance of water scarcity across England (Walker 2014 drawing on Jessop 2000, 2003). The lasting legacies of these historical forms of governance—and the conceptualization(s) of resilience that they push forward and promote regarding infrastructures, nature and people—can be witnessed throughout the events narrated in this chapter.

The Somerset case study—which reflects upon the policy aftermath of a period of drought and flooding between 2010 and 2014—is an insight into how these (meta) governance failures affect multiple areas of water management across England. The fragmented nature of the English water sector splits multiple responsibilities for different aspects across multiple actors (and due to the nature of the Levels and Moors the water management in Somerset is particularly unique and complex). The assessment of the governance context in Somerset has shown that the long-term adaptation plans, and crisis management strategies and instruments that are emerging in each of these boundaries of responsibilities for water management are strong and becoming increasingly clear (e.g. Environment Agency 2015a). However, it is how they unfold in locations and times of crisis that actually reflect the entrenched meta-governance failures. Such emergency events often lead to calls to renationalize the water services in England (e.g. Clark 2012; Gaines 2013)—stemming from a range of critiques of the neoliberalization of water sectors internationally (e.g. Bakker 2013; de Gouvella and Scott 2012; Hall et al. 2013). However, this is the governance system that has been inherited, and that is still emerging (Walker 2014) in events such as those described in this chapter. The events in Somerset reflect a deeper political failure to maintain strategies and instruments that support water management both directions of climate change extremes (drought and flood), and a failure to dovetail adaptation into connected forms of policy and planning (such as land use planning).

A discussion of the types of solutions and actions that could proactively deal with both drought and flooding is needed, at all political levels. In the aftermath of the flooding it was difficult to see how such a measured debate could in fact be initiated—after all these were lives and livelihoods that were devastated by the flooding and many are still involved in the necessary long-term emotional and physical work implicated in recovery from such extremes events (Medd et al. 2014; Whittle et al. 2011). Discussing 'drought and water scarcity' at such times could be seen to be highly political and highly insensitive to the lived experiences of flooding recovery (Medd et al. 2014). After all, water is materially and socially a highly emotive subject—whether through its overabundance, its lack, or when polluted (Sultana 2011). As can be seen in the Somerset case, the use, control and conflicts around water shape peoples everyday experiences with water (Sultana 2011). Water in complex landscapes such as Somerset have substantial emotive aspects, as discussion of its control and use intersects with experiences of place, livelihoods, and social, economic, political and environmental futures. It is these futures that are directly being intervened with in both longer term planning (such as water resource management plans, drought plans), and emergency management and policies that emerge during and after periods of crisis. Whittle et al. (2012) have captured strongly the emotional work that occurs simultaneously to the practical work in restoring built and natural environments after a flooding event. Clearly then the discussion of the future of water scarcity for an area such as Somerset at a time when the whole region is concentrating on recovering from a period of water abundance is an emotive and contested conversation. Nonetheless, it is a fundamentally important one.

Despite these restrictions in this broader governance system, the pilot measures initiated by a number of stakeholders in the Somerset region (cp. Sect. 5.3) demonstrate a positive example of the sorts of water management activities that can bridge across the policy silos of flood and drought even in a period of 'disaster recovery'. Far from just satisfying certain WFD requirements for participation and catchment management, these initiatives reflect concrete attempts to change the *experiences* of stakeholders and engagements with the breadth of water management issues facing their region now and into the future. These are highly politicised and emotive processes—they tap into conversations about 'what the Somerset Levels and Moors are for'; highlighting conflicts between protection of people, agriculture and nature; and arguably reflect the ongoing entrenchment of neoliberal and meta-governance failures in relations to drought and water scarcity across England (Walker 2014). As Clout (2014) reflects, such conversations about the purpose of the Levels—whether to maintain agriculture or support nature conservation—are not new and they have probably been happening since the middle ages, when the area was first drained and developed.

Within this project and chapter, we can neither prescribe what the future of the Levels should be or what activities need to take place in order to continue opening up conversations about ways to bridge silos within water management activities within the region. However, we do call on leaders and stakeholders in the Somerset region to continue collaborative processes of water governance across the widest possible range of stakeholders. This will ensure that a diversity of views from these stakeholders are catalogued, when adaptation policies are developed across water policy domains. Hopefully, in this way broader meta-governance failures—which often entrench siloed conceptualizations of water systems resilience—can then be avoided. At a national level, there needs to be a greater consideration of the ways to overcome these meta-governance failures which is currently limiting the regional resilience of vulnerable rural and urban water catchments across England to future climate changes.

References

Adey P, Anderson B (2011) Event and anticipation: UK Civil contingencies and the space-times of decision. Environ Plann A 43:2878–2899

Association of Drainage Authorities (ADA) (n.d.) An introduction to internal drain age boards (IDBs). Association of drainage authorities UK. http://www.ada.org.uk/downloads/publications/IDBs-An-Introduction-web.pdf. Accessed 16 Dec 2015

Bakker KJ (2003) An uncooperative commodity. Privatizing water in England and Wales, 1st edn. Oxford University Press, New York

Bakker KJ (2005) Neoliberalizing nature? Market environmentalism in water supply in England and Wales. Ann Assoc Am Geogr 95(3):542–565

Bakker K (2013) Neoliberal versus postneoliberal water. Geographies of privatization and resistance. Ann Assoc Am Geogr 103(2):253–260

BBC (2014) Somerset flood crisis: how the story unfolded. http://www.bbc.co.uk/news/uk-england-somerset-26157538. Online 19 Mar 2014 ay 2012

Browne AL (2014) Fast water versus slow water. Fragmentation in adaptation and resilience to flooding and water scarcity. http://blog.policy.manchester.ac.uk/featured/2014/02/fast-water-versus-slow-water-fragmentation-in-adaptation-and-resilience-to-flooding-and-water-scarcity. Policy@Manchester Blog posted 14 Feb 2014

Browne AL, de Boer C, La Jeunesse I, Stein U (2015) Assessment of the governance context of drought and water scarcity policy for Somerset, UK. Final report application of the governance assessment tool for water scarcity and drought adaptation. The DROP Project, University of Manchester, UK

Butler C, Walker-Springett K (2015) People and politics in the aftermaths of floods. [Blog]. http://geography.exeter.ac.uk/media/universityofexeter/schoolofgeography/winterfloods/documents/Blog_2015.pdf. Accessed 30 Jul 2015

Cameron D (2014) The Somerset levels and moors flood action plan. A 20 year plan for a sustainable future "We cannot let this happen again". [13 Feb]. HM Government, UK. http://somersetnewsroom.files.wordpress.com/2014/03/20yearactionplanfull3.pdf. Accessed 21 Oct 2015

Capstick SB, Demski CC, Sposato RG, Pidgeon NF, Spence A, Corner AJ (2015) Public perception of climate change in Britain following the winter 2013/2014 flooding. Cardiff University, Cardiff. http://orca.cf.ac.uk/68062/. Accessed 21 Oct 2015

Clark N (2012) Renationalise English water [released online 31st January 2012] http://www.
    theguardian.com/commentisfree/2012/jan/31/renationalise-english-water. Accessed 30 Jul
    2015
Clout H (2014) Reflections on the draining of the Somerset levels. Geogr J 180(4):338–341
Committee on Climate Change Adaptation (2014) Policy note. Flood and coastal erosion risk
    management spending [2014-01-21]. Adaptation sub-committee Secretariat. https://www.
    theccc.org.uk/wp-content/uploads/2014/01/2014-01-21-ASC-Policy-Note-flood-defence-
    spending-FINAL.pdf. Accessed 22 Oct 2015
Defra (2013) Catchment based approach. Improving the quality of our water environment A policy
    framework to encourage the wider adoption of an integrated catchment based approach to
    improving the quality of our water environment [PB 13934]. Defra, UK. https://www.gov.uk/
    government/uploads/system/uploads/attachment_data/file/204231/pb13934-water-
    environment-catchment-based-approach.pdf. Accessed 19 Oct 2015
De Gouvello B, Scott CA (2012) Has water privatization peaked? The future of public water
    governance. Water Inter 37(2):87–90
Environment Agency (2012a) South west drought plan. Environment Agency, Bristol, UK. https://
    www.gov.uk/government/uploads/system/uploads/attachment_data/file/292773/
    gesw0112bvyh-e-e.pdf. Accessed 19 Oct 2015
Environment Agency (2012b) Head office drought plan. Environment Agency, Bristol, UK.
    https://www.gov.uk/government/uploads/system/uploads/attachment_data/file/297211/
    geho0112bway-e-e.pdf. Accessed 20 Oct 2015
Environment Agency (2014) Effectiveness of additional dredging: part of the 20 year flood action
    plan. Environment Agency, Bristol, UK. https://www.gov.uk/government/uploads/system/
    uploads/attachment_data/file/392471/The_effectiveness_of_dredging_elsewhere.pdf. Accessed
    19 Oct 2015
Environment Agency (2015a) Drought response: our framework for England. HM Government,
    UK.   https://www.gov.uk/government/uploads/system/uploads/attachment_data/file/440728/
    National_Drought_Framework.pdf. Accessed 28 Jul 2015
Environment Agency (2015b) Somerset levels and moors. Reducing the risk of flooding [notice
    version updated 14 January 2015]. HM Government, UK. https://www.gov.uk/government/
    publications/somerset-levels-and-moors-reducing-the-risk-of-flooding/somerset-levels-and-
    moors-reducing-the-risk-of-flooding. Accessed 20 Oct 2015
Fenn C, Wilby R (2011) Smarter licensing to reduce damage abstraction from environmentally
    fragile rivers with minimum possible impact on water resource yield. River Itchen case study.
    Report. WWF-UK, London
Gaines M (2013) Rentationalising water industry should be a priority for Labour. Printed online 18
    Jul 2013. http://wwtonline.co.uk/news/renationalising-water-industry-should-be-a-priority-for-
    labour#.VboOEPlViko. Accessed 30 Jul 2015
Gilbert E (2012) The militarization of climate change. ACME. Int e-J Crit Geogr 11(1):1–14
HM Government (2008) Future water. The government's water strategy for England. Report Cm
    7319, HM Government, London. https://www.gov.uk/government/uploads/system/uploads/
    attachment_data/file/69346/pb13562-future-water-080204.pdf. Accessed 22 Oct 2015
HM Government (2011) Water for Life. Report CM 8230 presented to parliament by the secretary of
    state for environment food and rural affairs by command of her majesty, HM Government, London.
    http://www.official-documents.gov.uk/document/cm82/8230/8230.pdf. Accessed 20 Oct 2015
HM Government (2012) Draft water bill. Report CM 8375 presented to parliament by the secretary of
    state for environment food and rural affairs by command of her majesty, HM Government, London.
    http://www.official-documents.gov.uk/document/cm83/8375/8375.pdf. Accessed 20 Oct 2015
HM Government (2013) The national adaptation programme. Making the country resilient to a
    changing climate. Ref: PB13942. Report presented to parliament pursuant to Section 58 of the
    climate change act 2008, HM Government, UK. https://www.gov.uk/government/publications/
    adapting-to-climate-change-national-adaptation-programme. Accessed 20 Oct 2015
Hall D, Lobina E, Terhorst P (2013) Re-municipalisation in the early twenty-first century. Water in
    France and energy in Germany International. Rev Appl Econ 27(2):193–214

Hope C (2012) Thames water accused of 'mismanagement' by closing two dozen reservoirs. The telegraph, 30 Apr 2012. http://www.telegraph.co.uk/news/earth/drought/9236909/Thames-Water-accused-of-mismanagement-by-closing-two-dozen-reservoirs.html. Accessed 30 Jul 2015

Jessop B (2000) Governance failure. In: Stoker G (ed) The new politics of British local governance. Macmillan, Basingstoke, pp 11–32

Jessop B (2003) Governance and meta-governance: On reflexivity requisite variety and requisite irony. In: Bang H (ed) Governance as social and political communication. Manchester University Press, Manchester, pp 101–116

Lever A-M (2012) Redefine concept of drought, environment agency urges. BBC science and environment. http://www.bbc.co.uk/news/science-environment-18150100. Accessed 22 May 2012

Maloney WA, Richardson J (1994) Water policy-making in England and Wales. Policy communities under pressure? Environ Polit 3(4):110–138

McEwen L, Jones O, Robertson I (2014) 'A glorious time?' Some reflections on flooding in the Somerset Levels. Geogr J 180(4):326–337

Medd W, Marvin S (2008) Making water work. Intermediating between regional strategy and local practice. Environ Plann D: Soc Space 26(2):280–299

Medd W, Deeming H, Walker G, Whittle R, Mort M, Twigger-Ross C, Walker M, Watson N, Kashefi E (2014) The flood recovery gap. A real-time study of local recovery following the floods of June 2007 in Hull, North East England. J Flood Risk Manag

Natural England (2013) National character area profile 142. Somerset levels and moors. file:///C:/Users/MBRXSAB8/Downloads/NE451.pdf. Accessed 29 Jul 2015

Natural England (2014) An assessment of the effects of the 2013–14 flooding on the wildlife and habitats of the Somerset levels and moors. Natural England, England. http://www.sanhs.org/Documents/FloodingNE.pdf. Accessed 29 Jul 2015

Rahiz M, New M (2012) Spatial coherence of meteorological drought in the UK since 1914. Area 44(4):400–410

Rahiz M, New M (2013) 21st Century Drought Scenarios for the UK. Water Resour Manage 27 (4):1039–1061

Somerset Drainage Boards Consortium (SDBC) (2014) About us [webpage] http://www.somersetdrainageboards.gov.uk/about-us/. Accessed 14 Dec 2015

Sultana F (2011) Suffering for water suffering from water. Emotional geographies of resource access control and conflict. Geoforum 42:163–172

Taylor V, Chappells H, Medd W, Trentmann F (2009) Drought is normal. The socio-technical evolution of drought and water demand in England and Wales 1893-2006. J Hist Geogr 35 (3):568–591

Thorne C (2014) Geographies of flooding. Geogr J 180(4):297–309

van Loon AF, van Lanen HAJ (2013) Making the distinction between water scarcity and drought using an observation-modeling framework. Water Resour Res 49(3):1483–1502

Wade SD, Rance J, Reynard N (2013) The UK climate change risk assessment 2012: assessing the impacts on water resources to inform policy makers. Water Resour Manage 27:1085–1109

Walker G (2014) Water scarcity in England and Wales as a failure of (meta) governance. Water Altern 7(2):388–413

Water UK (2015) Price reviews. http://www.water.org.uk/price-reviews. Accessed 12 Dec 2015

Waterwise (2013) Water efficiency and drought communications report. Waterwise/Water UK/WWF-UK, London. http://www.waterwise.org.uk/data/2013_Waterwise_Drought_Report.pdf. Accessed 19 Oct 2015

Wharfe P (2015) Droughts and floods. Towards a more holistic approach. Releasing the full value of UK Research. UK Water Partnership, Research and Innovation Group, London. http://www.theukwaterpartnership.org/#projects-and-publications. Accessed 20 Oct 2015

Whittle R, Walker M, Medd W (2011) Suitcases storyboards and newsround. Exploring impact and dissemination in Hull. Area 43(4):477–487

Whittle R, Walker M, Medd W, Mort M (2012) Flood of emotions. Emotional work and long-term disaster recovery. Emot Space Soc 5(1):60–69

Chapter 6
The Governance Context of Drought Policy and Pilot Measures for the Arzal Dam and Reservoir, Vilaine Catchment, Brittany, France

**Isabelle La Jeunesse, Corinne Larrue, Carina Furusho,
Maria-Helena Ramos, Alison Browne, Cheryl de Boer,
Rodrigo Vidaurre, Louise Crochemore, Jean-Pierre Arrondeau
and Aldo Penasso**

6.1 Introduction

This chapter presents an analysis of the drought adaptation governance of the Vilaine catchment in the Brittany region in France and, more specifically, of the Arzal dam and reservoir located at the outlet of the river. Accordingly, the analysis focuses on the lower part of the Vilaine catchment, where two pilot studies were conducted during the DROP project. The material for the analysis was collected during two field visits. The first visit occurred from 16 to 18 September 2013, and the second from 16 to 18 June 2014, during which the Governance Team (GT) met stakeholders, managers and representatives of the relevant local action groups. The analysis is also based on several documents provided by the *Institution d'Aménagement de la Vilaine* (IAV),

I. La Jeunesse (✉) · C. Larrue
University of Tours, 33, allée Ferdinand de Lesseps, 37204 Tours, CX 3, France
e-mail: isabelle.lajeunesse@univ-tours.fr

C. Furusho · M.-H. Ramos · L. Crochemore
ISTEA, Rue Pierre-Gilles de Gennes 1, Antony, France

A. Browne
University of Manchester, Manchester M13 9PL, UK

C. de Boer
Faculty of Geoinformation Science and Earth Observation, ITC, University of Twente,
Drienerweg 99, 7522ES Enschede, Netherlands

R. Vidaurre
Ecologic Institute, Pfalzburger Strasse 43/44, 10717 Berlin, Germany

J.-P. Arrondeau · A. Penasso
Institution d'Aménagement de la Vilaine, Pfalzburger Strasse 43/44, La Roche Bernard,
France

© The Author(s) 2016
H. Bressers et al. (eds.), *Governance for Drought Resilience*,
DOI 10.1007/978-3-319-29671-5_6

one practice partner of the DROP project responsible for the management of the Arzal dam and its reservoir. IAV also plays a key role in supporting water management at the catchment scale of the Vilaine River.

After a presentation of the French drought context (Sect. 6.2), including the description of the water management and drought adaptation strategies adopted in France, this chapter describes the main water use management issues for the Arzal dam and reservoir (Sect. 6.3). Section 6.4 is dedicated to the interpretation of the results of the drought governance analysis which was supported by the implementation of the drought governance assessment tool (GAT). Finally, in Sect. 6.5, some recommendations are proposed regarding possible measures to improve local drought adaptation strategies.

6.2 National Drought Governance Context

6.2.1 Some Past Drought Events and Consequences on Water Policy

Periods of drought experienced in France since the 1970s (Corti et al. 2009) have usually led to changes in national policies or in drought adaptation measures. The main characteristics and impacts of four major events are presented in Table 6.1.

The drought of 1976 was especially severe in the northern half of the country, also affecting other parts of northwestern Europe (Le Roy Ladurie et al. 2011). Given the strong impacts of this drought on the agricultural sector, the French government provided assistance to farmers with funds collected from the creation of a new "drought tax". The drought of 1989 also saw a large deficit in soil moisture. Although a drinking water supply could be ensured in most cases, measures concerning prohibitions or restrictions on certain water uses had to be adopted. The situation highlighted the general lack of regulation in France: the inability to quantify water abstraction, the absence of groundwater abstraction control and the lack of regulatory tools for allowing the authorities to allocate the remaining resources. In this context, the revision of the 1964 Water Law in France was accelerated, with a new revision being passed in 1992. Under the provisions of the current Environmental Code, the Prefects (the State representatives in the French decentralization process initiated in the 1980s) can take exceptional measures to address a possible lack of water resources. The drought of 2003 and its exceptional heat wave led to the creation of a Drought Action Plan (*Plan d'Action Sécheresse*), aiming to reconcile different water uses while preserving the quality of aquatic environments. In addition, a National Drought Committee (*Comité Sécheresse*) was created to coordinate water uses during drought crises. This Committee meets when needed depending on the hydrological situation immediately before or during a dry period. The Low Flow Management Plan (*Plan de gestion des étiages*, PGE) and the Drought Action Plan were implemented as operational tools for the management of hydrological droughts. The drought of 2011

Table 6.1 Main characteristics of the European droughts of 1976, 1989, 2003 and 2011 and their main impacts on **French water policy**

Year	Spatial extent	Time period	General description	Main impacts in France (EDC 2013)
1976	Europe, more specifically northern Europe	Winter 1975 to summer 1976	Characterized by a very long period with precipitation deficits, from December 1975 until August 1976, associated with deficits of groundwater recharge. The northwestern part of France was particularly affected. Historic high temperatures were observed beginning with the end of June 1976, and the heat wave persisted for three weeks, up to mid-July (Brochet 1977; Zaidman et al. 2002; Vidal et al. 2010)	• Human losses in France because of the heat wave • Severe loss of spring crops • Death of fish and livestock • 1.7 MT of straw transferred (Brochet 1977) • Restriction on drinking water supply • Wildfires • High concentration of pollutants in rivers • Reduction of hydropower and nuclear power production • **Drought tax** to FFRANCS 2.2 billion
1989	Most European countries	Summer to autumn 1989, with impacts lasting until end of summer 1990	Characterized by multiple short drought events (Zaidman et al. 2002) and extremely dry soils for several months. The event was caused by low precipitation and high temperatures for a prolonged period from autumn 1988 to autumn 1989. Dry soils, low snowpack, low groundwater levels and severe low flows were observed during spring and summer 1989 (Mérillon and Chaperon 1990). The low groundwater levels after summer 1989 led to dry summers in 1990 and 1991, despite average meteorological conditions (https://erisk.ccr.fr/)	• Loss of crops • High concentration of pollutants in rivers • Death of fish • Wildfires • Reduction of hydropower and nuclear power production • **Temporary water ban** through revision of the Water Law (passed in 1992)

(continued)

Table 6.1 (continued)

Year	Spatial extent	Time period	General description	Main impacts in France (EDC 2013)
2003	Europe, more specifically west Europe	Summer 2003	Characterized by a wide spatial extent and an exceptional 15-day heat wave (Poumadère et al. 2005). In France, all regions were affected. Rainfall deficits from February to September were approximately 20–50 %. A soil moisture deficit was observed over France, resulting in several regions experiencing by soil subsidence for the first time. Compared with the long drought of 1989, the event was relatively short (Vidal et al. 2010)	• 14,800 human losses in France (Robine et al. 2007; Pirard et al. 2005) • Important loss of crops (UNEP 2004) • Death of fish and livestock, wildfires and insect invasions • Decrease in nuclear power production (UNEP 2004) • Water use restrictions in 75 % of the French departments • Natural Disaster Declaration for 4400 municipalities • Drought-induced soil subsidence (Corti et al. 2009) • Economic costs estimated to EUR 1.1 Billion • **Drought Action Plan and National Drought Committee**
2011	Most European countries	Spring 2011	Characterized by a dry 2010–2011 winter and record high temperatures. The spring average precipitation deficit was over 40 %, and the average spring temperature was 2.5 °C above normal (1959–2011 https://erisk.ccr.fr/). In France, the 2011 drought particularly affected the southwestern regions	• Loss of crops • Difficulties with cattle feeding, farmers having to sell their animals • Wildfires • Water use restrictions in 75 % of the French departments • Historic deficit of hydropower production • **National guarantee funds through agricultural disaster** • **Implementation of PROPLUVIA tool to manage water bans** (http://propluvia. developpement-durable.gouv.fr/propluvia/faces/index.jsp)

strongly impacted the agricultural, animal farming and energy sectors. Its impact on agriculture was comparable, in its severity, to the 1976 drought (Vidal et al. 2008, 2010). To cope with this drought, the French Ministry of Agriculture provided several hundred million euros in financial assistance to farmers under the national guarantee fund, specifically dedicated to agricultural disasters.

6.2.2 Water Management in France

France is divided into administrative regions and departments. For water, the territory is divided into twelve hydrographic districts each having its own river basin committee (*Comité de Bassin*) composed of representatives of the local authorities (40 %), water users (40 %) and State representatives (20 %). The committee establishes guidelines according to the European Water Framework Directive (WFD) and national water policies (Water Law since 1992, updated in 2006 under the LEMA law name). The guidelines are driven by the river basin management plan (SDAGE, *Schéma Directeur d'Aménagement et de Gestion des Eaux*), the six-year catchment-wide development and environmental plan promoted by water agencies (*Agences de l'eau*) (Fig. 6.1).

At the national level, the central government is in charge of all regulation issues (e.g. authorizations for water abstractions and pollution, monitoring and control of implementation by water users). The Water Management Direction and the French Ministry of Ecology, Sustainable Development and Energy are responsible for defining water management policies, including drought control and drought management plans for areas sensitive to droughts.

Within the territory, the decision and implementation of water regulation and drought control policies relies strictly on the Prefects of the departments, who have represented the central government at this level since 1982. Prefects are empowered to issue, implement, coordinate and control general and individual bylaws (*Arrêtés*) on water use, and, particularly, on drought control, through water bans, determined with the local support of mayors of municipalities and regional ministry entities (DREAL).

As concrete operational management is conducted at a river basin scale, local initiatives have an increasing role in water management in France. While the water agencies manage very large river basins, local issues are managed by groups of local authorities around a river section—the public body called EPTB (*Etablissement Public Territorial de Bassin*), which is similar to the 'waterboards' in the Netherlands. An EPTB, status established in 1997, can act in three areas: hydraulics (low-flow management, flood prevention, water production), environment (actions for migratory fish, maintenance of the banks, basin observatories) and local development (actions towards natural and cultural heritage). EPTBs have the status of interdepartmental institutions or joint local authorities. Their funding is provided by their own members. EPTBs are usually the coordinators of the SAGE (*Schéma d'Aménagement et de Gestion de l'Eau*), which is the water management plan at the sub-catchment level. SAGE is a non-constrained structure that acts

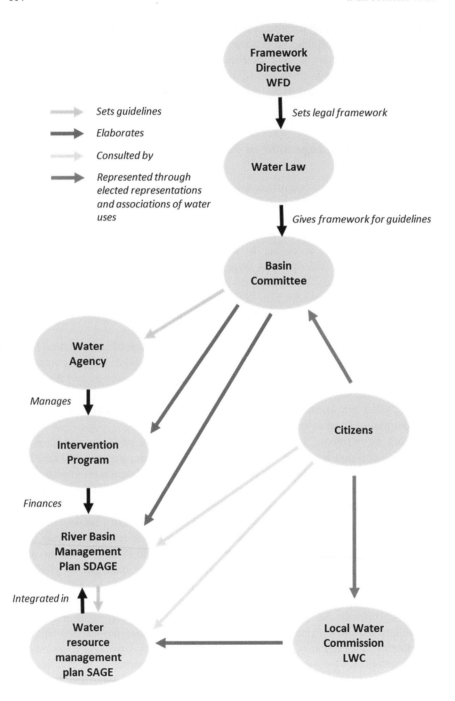

Fig. 6.1 Water management in France

through a Local Water Commission (LWC, or CLE for *Commission Locale de l'Eau* in French) composed of State representatives (25 %), local authorities (50 %) and users (25 %). The members of the LWC are, thus, the main actors in the development of the SAGE.

Since 2006, the SAGE includes a plan for sustainable management of water resources and aquatic environments within a new document, the PAGD (*Plan d'Aménagement et de Gestion Durable de la ressource en eau et des milieux aquatiques*), with a plan of action aiming at achieving good ecological status for the sub-river basin. Urban planning documents must also comply with the objectives of water resource protection defined by the SAGE.

6.2.3 Drought Adaptation in France

In response to the frequency of water scarcity situations that occurred during the drought episodes observed at the end of the twentieth century, as described in Sect. 6.2.1, the French public authorities have gradually developed drought control policies and short-term mitigation action plans. Drought control management in France is now part of the legal frameworks and procedures governing general national water management. Measures to address droughts have two primary basic aspects: crisis management through emergency actions and mean to long-term quantitative/qualitative water resources management. The former involves measures set up within the framework of the river basin management plans (SDAGE), while the latter is more related to the national plan that copes with climate issues, which is implemented through the regional climate change action plans (named SRCAE, for *Schéma regional climat, air, énergie*, and meaning "Regional Scheme for Climate, Air and Energy").

6.2.3.1 Emergency Actions

Temporary periods of drought and water scarcity in France are generally managed through emergency actions, such as water use bans on non-priority uses, which are under the authority of the State. Bylaws imposing individual and collective restrictions and bans on water use are temporary. Drought control measures must be proportional to the water scarcity level and adapted to the required mitigation actions. In 2005, methodological guidelines to be followed in the case of a drought were published by the national government. This document recommends a progressive phasing of four levels of alert (monitoring/alert/crisis/critical crisis), which are adapted to the local water resource situation and historical background. It can be supported by the online cartographic tool PROPLUVIA.[1]

---

[1]http://propluvia.developpement-durable.gouv.fr/propluvia/faces/index.jsp.

With the Drought Action Plan that was officially implemented in 2004, each region is to maintain updated information on the regional context (geographical features, climatology, hydrological aspects, etc.), the actors involved in decision-making, the areas sensitive to droughts, the different river discharge values that define critical low-flow thresholds to support the characterization of four levels of alert, the mayor's role in the process, the past water abstraction restrictions and the methods of communication and public information, all this in collaboration with the SDAGE river basin management plan. The objective is to improve the water resource monitoring systems and the coherence of bans on water uses, as well as to enhance the coordination of actions at the catchment scale, especially between upstream and downstream water users from different administrative units.

Since January 2012, a national observatory whose purpose is to study low flows (*Observatoire National Des Etiages*, ONDE) has been implemented by the National Agency for Water and Aquatic Environments[2] (ONEMA). This observatory has a dual purpose: building a solid database of knowledge on summer low flows and developing tools to manage drought crises. In France, the south and southwestern regions are currently the most sensitive regions to droughts, while the north and northwestern regions generally do not require long-lasting water bans. However, even in a catchment located in northwestern France, as is the case for the Vilaine catchment, although it rarely suffers from severe droughts compared to what is observed in the southern part of the country, it can happen that water managers have to face episodic problems of imbalance between water availability and demand. This can create situations where it becomes difficult to satisfy all uses and where tensions may occur.

6.2.3.2 National Plan to Cope with Climate Change

Within the recent national climate adaptation action plan driven by the Ministry for the Environment (*Plan national d'adaptation de la France aux effets du changement climatique* 2011–2015), a set of actions regarding drought and water scarcity is proposed. The following hierarchical priorities are considered regarding the types of measures to be taken or planned:

- No-regret measures, which are beneficial even without climate change: in the field of water, for instance, it includes the promotion of water saving;
- Reversible measures, which give leeway of action for adaptation: in the field of water, measures such as the revision of levels of reference for flood prevention or low-flow control;
- Long-term measures, such as the integration of climate change in the long-term plans for drought adaptation;
- Measures that can be revised periodically as improved knowledge becomes available.

---

[2]A national public agency whose mission is to restore the healthy ecological status of water and aquatic environments, within the goal set by the European Water Framework Directive.

Regional responsibility for climate change adaptation lies within the Regional Scheme for Climate, Air and Energy (SRCAE) and the Climate-Energy Plans (PCET), which are currently being developed at the local level (i.e. for municipalities with 50,000 inhabitants or more).

6.3 Geo-Hydro Context, Drought Policy Focus and Measures Taken in the Vilaine

6.3.1 The Vilaine River

6.3.1.1 Hydrological Description

The Vilaine River is a river nearly 230 km long flowing into the Atlantic Ocean. Its catchment area, located in the Loire-Bretagne river basin (Fig. 6.2), is 10,400 km^2 in size. In 2012, the catchment area of the Vilaine had 1,260,000 inhabitants. In approximately 1960, actions were undertaken to restore navigability on the lower Vilaine River and boost the economic development of its main city, Redon. The aim was to stimulate local industry while reducing flooding and desalinating the marsh so it could be devoted to agricultural activities.

The catchment receives an annual average of 700–800 mm/year of precipitation. Mean daily flows are higher from December to March, with an average of 150 m^3/s,

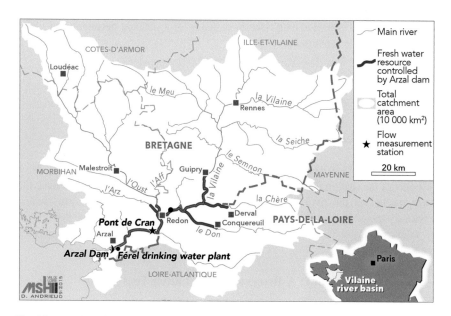

Fig. 6.2 Location of the Vilaine catchment, the Arzal dam and the zone of influence of the dam in the upper part of the riverbed

while they are significantly lower from July to October, with an average of 15 m³/s. In the Vilaine catchment, the low-flow period runs from May/June to October.

6.3.1.2 Drought Threats and Water Scarcity in the Vilaine Catchment

Traditionally considered to enjoy water in abundance, the northwestern regions of France do not always have drought policy and management plans formulated at the river basin scale. In the Vilaine River, however, provisions have been taken to create frameworks for some main water-related issues, notably for drinking water distribution and low-flow management in some sensitive parts of the catchment. In fact, the Vilaine catchment is heterogeneous regarding drought threats. The eastern part of the catchment is classified as more sensitive to water scarcity because of its exposure and vulnerability, as irrigated crops are rooted in these areas. Since 2011 (Regional Decree of 24 January 2011), the entire Vilaine catchment is no longer classified as a water-sensitive area (in France, ZRE, *zone de repartition des eaux*). In France, a 'water-sensitive area' refers to an area where water resources are insufficient compared to water uses, and the area is thus considered to require special protection regarding low-flow water levels. In water-sensitive areas, abstractions other than for drinking water are limited between 1 April and 30 October at their current level. Small dams may be permitted only if they have no impact on low-flow rates. Because the Vilaine catchment is no longer considered a 'water-sensitive area', it is no longer possible to constrain water uses permanently in the Vilaine area (although water use restrictions may be put in place during dry years, i.e. in below mean situations). Recent water management measures in the Vilaine claim that the catchment should again be registered as a 'water sensitive area' (see Sect. 6.3.3 for a detailed description of water management in the catchment).

6.3.2 The Arzal Dam

6.3.2.1 One Initial Objective: Regulating the Risk of Floods

The Arzal dam (Fig. 6.3) is located at the mouth of the Vilaine catchment, just above the outlet of the river to the Atlantic Ocean (Fig. 6.2). It was built in 1970 to isolate the lower reaches of the river basin from the ocean and block the tidal flow, which previously travelled inland as far as the city of Redon and could aggravate flooding in the valley when floods in the Vilaine River occurred simultaneously with a high tide.

6.3.2.2 An Opportunity: A Reservoir with Multiple Uses

In addition to regulating the risk of floods, the dam also allowed the creation of a reservoir of 50 Mm³ of freshwater, stretching up to 80 km. A water supply

Fig. 6.3 The Arzal dam (*Source* IAV, with authorization)

Fig. 6.4 The drinking water production plant (Drezet at Férel) that treats the water from the Arzal reservoir (*Source* IAV, with authorization)

treatment plant (Drezet at Férel, Fig. 6.4) was inaugurated in 1972, with a production capacity of 30,000 m^3 of drinking water per day, corresponding to an instantaneous pumping rate in the reservoir of 1600 m^3/h. Today, its maximum production capacity has been increased to 90,000 m^3 per day, corresponding to an instantaneous pumping rate in the reservoir of 4800 m^3/h. The water plant supplies 15–20 Mm^3 of freshwater annually. Up to 1 million inhabitants are supplied with drinking water by the Arzal reservoir during the summer season.

Fig. 6.5 A yachting marina in the Vilaine River before the Arzal dam (*Source* La Jeunesse I., with authorization)

The Arzal dam also encouraged the development of new recreational activities such as yachting. Sailing is now highly developed in the area (Fig. 6.5), with an average of 18,000 boats crossing the dam per year, and approximately 85–90 % of boat crossings occurring during May to October, thus, during the low-flow period (see Sect. 6.3.1). Sailing and fish passages have, however, an influence on the quality of the water in the reservoir. Freshwater quality in the reservoir can be

Fig. 6.6 The lock of the Arzal dam (*Source* IAV, with authorization)

affected primarily by salt water intrusions. They occur primarily when the lock is operated for navigation (Fig. 6.6). They are usually handled by a system of syphons that automatically evacuates salt water from the reservoir to the estuary. Because the most significant inputs of salt water to the reservoir occur when boats pass through the existing lock of the dam, navigation can be restricted during the summer. Management rules with restrictions on the time schedule and number of lock openings can thus generate conflicts, especially during the summer high season for recreational sailing.

Today, the Arzal dam consists of five elements: an unsinkable dike, a set of five sluices, a lock, a syphon and a fish passage.

6.3.2.3 Pilot Measures Implemented Within the DROP Project

The Arzal dam and its reservoir are managed by IAV, the practice partner of the DROP project for the Vilaine case study. IAV is an EPTB (see Sect. 6.2.1), which means that it is a French public organization for the cooperation of local authorities (regions, departments, municipalities, etc.) towards the development and management of rivers in the geographical context of a river catchment or sub-catchment. IAV faces challenges related to salt water intrusion and reservoir management during the low-flow season, when both water quantity and water quality constraints apply on the management of the dam. As mentioned earlier, salt intrusions in the reservoir mainly occur when boats pass through the lock at the dam. When water inflows tend to be the lowest, recreational activities, including sailing, are generally at their highest level (because it involves the summer period), and lead to a peak of salt water intrusions in the reservoir, which can affect freshwater quality and, thus, quantity. To prevent salt intrusions, syphons have been installed upstream of the dam to pump the contaminated water from the reservoir back to the Atlantic Ocean. However, this system causes a significant loss of resources: approximately, 300,000 m^3 of freshwater are evacuated every day from the reservoir, compared to an average of only 10,000 m^3 of salt intrusions. In addition, the syphons are not 100 % efficient: some brackish water still enters the reservoir, and salinity peaks are regularly observed in late summer each year. Moreover, this system leads to losses of substantial quantities of freshwater, which, during prolonged periods of drought, may aggravate the problem of multiple uses of the reservoir and impact the freshwater supply.

Currently, the only way to limit salt water intrusions and, consequently, water losses from pumping is to place restrictions on the use of the lock in summer. The two principal strategic management objectives during low-flow periods are (i) to ensure a level in the Arzal reservoir that is adequate to permit all uses and (ii) to keep saltwater out of the Arzal reservoir as much as possible to preserve freshwater quality and guarantee the drinking water supply. However, these management objectives are not always satisfactory because they may result in restrictions in the use of water, which may not be fully accepted by other water users in the catchment: a preventive rise in water levels to supply summer water needs may result in flooding the wetlands during the hay harvest, forcing IAV to pay financial compensation to farmers; moreover, restrictions in the use of the lock at the dam during summer periods are seen by boat owners as an infringement of their freedom of navigation, a practice especially resented by boaters, as they invest money to maintain this activity.

This context has prompted IAV, over a long period of time, to consider the implementation of new solutions. The DROP project, comprising interactions with other practice partners managing freshwater reservoirs and with experts in governance analysis, was an opportunity to improve the reflections that were initiated in the course of optimizing the management of the reservoir during drought periods. Concretely, two main pilot measures were taken during the DROP project, a new lock and drought forecasting:

(a) The implementation of a new lock at the dam

IAV has worked on developing a new and innovative lock at the dam to prevent salt water intrusion when boats cross the dam. Significant efforts have been put into designing this new lock: preliminary and feasibility studies, 3D hydraulic modelling, and even a physical model at the 1/12 scale. Currently, all preliminary studies are finished, hydraulic and physical models have been calibrated, all simulations have been completed, design plans have been achieved, and a consolidated estimate of the cost of the project has been conducted.

(b) The development of drought forecasting and risk management tools

In parallel, IAV has been working with the IRSTEA research centre, which is also a partner of the DROP project, to develop a modelling system to forecast inflows to the reservoir during the low-flow season and help anticipate critical situations to ensure better drought risk management. The system incorporates weather information into a hydrological forecasting model and translates the results into a graphical representation of the drought risk. Future possible weather scenarios over the Vilaine catchment can thus be considered and transformed into river inflows all the way upstream to the dam. The graphical representation of the drought risk provides a visual assessment of the risk of being below given critical low-flow thresholds in the next weeks or months, both in terms of flow intensity and duration (i.e. mean flow and number of days below each critical threshold, respectively). This risk assessment visualization tool aims to help the managers of the dam in deciding whether to release water from the reservoir and on how to operate the corresponding dam components. It can be integrated into the various reservoir operations and management rules necessary to fulfil its multiple operational uses, connecting the utilities in a pre-operational framework. The tool is based on the development of a global forecasting chain, including the development of weather scenarios combining a short-term meteorological forecast (9 days), a long-term meteorological forecast (3 months) and an analysis of past events over the last 50 years. This pilot answers to the needs of drought alert tools, as also promoted by the national methodological guidelines presented earlier (Sect. 6.2.3).

6.3.3   Water Management in the Vilaine Catchment

6.3.3.1   The Main Instrument Devoted to Water Management in the Area: The SAGE Vilaine

A SAGE (Fig. 6.1, Sect. 6.2.2) is in force in the Vilaine catchment area. It is the largest in area of any SAGE in France. In terms of territorial governance, it involves two regions (Bretagne and Pays-de-la-Loire), six departments (Ille-et-Vilaine, Morbihan, Loire-Atlantique, Côtes d'Armor, Mayenne, Maine-et-Loire) and 534 municipalities. An initial water management plan was enacted in 2003. It was revised in 2015 and now includes a plan for sustainable management of water resources and aquatic environments (PAGD), as required by the water law of 2006.

In general, issues regarding the management of the Arzal dam are closely related to the dynamics of the water levels in the reservoir and in the river reaches influenced by the reservoir. This linkage plays an important role in the water governance of the whole Vilaine catchment and, particularly, of the lower part of the river basin. Three specific management issues are (i) the siltation problem in the estuary, (ii) the management of agriculture and hunting in wetlands and (iii) the salinization of freshwater related to the passage of the dam by boats. Following the conflicts induced by the management of the dam, IAV proposed to build new committees to represent all stakeholders concerned by these issues, with two initiated directly by the SAGE Vilaine: the estuary committee and the Natura 2000 committee. The objectives of these committees are to enhance solidarity among water users sharing the same resources and to improve coherence between general water resources management and drought control measures.

6.3.3.2   The Estuary Committee

Since the creation of the dam, the salinity gradient and the transfer of sediments in the river have been heavily modified. The estuary is facing rapid siltation, with no satisfactory solutions implemented at the moment. These ecological modifications have impacted fishing and shellfish farming both in the estuary and in the river. The rapid siltation prompted the creation of an Estuary Committee in 1999, which, as a consultative body, aims to specifically address the issue of the silting of the estuary in the Local Water Commission (LWC).

6.3.3.3   The Natura 2000 Committee

The marshes of Vilaine and Redon form nearly 10,000 ha of alluvial grasslands upstream from the Arzal dam. These marshes were, for the most part, the inner estuary of the Vilaine until the construction of the Arzal dams. These marshes belong to the European site network Natura 2000, which aims to preserve endangered biodiversity in Europe. As explained earlier, the regulation of the water level of the Arzal reservoir may induce flooding in the upper wetlands during the dry season. The Natura 2000 committee represents hunters, naturalists and some farmers with concerns about the periodicity of the flooding of those natural areas. The committee was created at the same period as the site network and contributes actively to the SAGE Vilaine.

6.4 Assessment of Drought Governance Qualities

In this section, we present our findings regarding the observations on the four qualities of the governance assessment tool, namely, extent, coherence, flexibility and intensity, assessed on the five governance dimensions of the matrix that forms the framework of the GAT (Bressers et al. 2013a, b, c).

To the greatest extent possible, this analysis distinguishes between (a) water management and governance as such and (b) governance related to drought adaptation. These elements do not have the same aims. Water management is part of an integrated system that is implemented in almost all decision-making bodies. Adaptation to drought is mainly restricted to water level regulation and the management of water crises during drought.

6.4.1 Extent: Large for Water Management and Limited for Drought Management

The **levels and scales** dimension has a truly supportive basis in the water management of the Vilaine catchment. The administrative management of the entire Vilaine catchment has a wider extension, with scales ranging from the European level at the top through the national, regional, departmental, and intercommunal levels and all the way down to the communal level. As for the SAGE Vilaine (Sect. 6.2.2) and the management of the Arzal dam and its reservoir by IAV, the involvement of representatives of each level is not the same: three departments (Morbihan, Ille-et-Vilaine, Loire-Atlantique, Fig. 6.2) are more involved in the decision-making process than the others. This can be explained by the fact that the other three departments within the Vilaine catchment (Côtes-d'Armor, Mayenne, Maine-et-Loire, Fig. 6.2) are less affected by the management of the Arzal dam. However, the whole catchment of the Vilaine is involved in the local water commission (LWC), representing six departments within two regions. Drought management relies on the same organization as that of water management, and, for that reason, the levels and scales that are potentially included are as large as the ones involving water management. However, water crisis management is situated at the national level.

Regarding **actors and networks**, users and managers have shown mutual trust thanks to frequent formal and informal meetings driven by IAV for discussing shared issues. There is also a strong awareness of the decision-making processes in water management. None of the interviewed actors expressed the feeling of being excluded from any decision-making process. IAV has thus built a strong network, involving the most relevant actors within the territory, and has become an institution with remarkable political involvement and influence. However, concerning drought management, we can observe that there is no network of actors specifically dedicated to address drought management at any scale except for crisis management at the national level, as mentioned earlier.

The LWC and the SAGE enable any issues related to water management to be addressed. As a result, the **problem perceptions and goal ambitions** can potentially have a broad extent provided that all issues are addressed. However, this is not yet the case. Before the DROP project, for instance, the impacts of climate change on drought frequency and intensity had never been considered a topic to be discussed by the LWC. In fact, drought is not a hazard mentioned frequently in discussions related to water management. Several considerations can explain this observation. First, the area is a wet region and, even though it has seen severe drought events in the past, they remain rare. Additionally, the several dams that exist along the Vilaine River, including the Arzal dam located at the outlet of the catchment, are already seen as infrastructures that tackle the problems of water scarcity. This is observed even though some eastern parts of the catchment, which are more concerned by intensive agriculture practices, are already experiencing water scarcity. Moreover, drinking water is the central issue for the manager of the Arzal dam. The perceptions of the impacts of climate change and its consequences on the frequency and severity of drought events are still largely relegated to the State administration at the national scale, and, locally, are mainly oriented towards crisis management.

The main **instruments** for regulating water use are integrated in the SAGE Vilaine with a general **strategy** concerning regulations, incentives and communication. However, and this is the case for all the country, even during drought periods, the price of drinking water cannot be considered an economic adjustment tool. This is because the national legislation in France ("drinking water pays for drinking water") imposes an independent budget. The price of water is fixed and indexed to the cost of its management. It cannot be used as a regulation instrument in situations of water scarcity. For drought concerns specifically, one can state that there is a restricted extension related to strategy and instruments. This can be explained by the drought adaptation strategy developed nationally. Since the frequency of drought events is not high in the Vilaine catchment, it is not classified as a quantitative 'water sensitive area', as mentioned earlier. For this reason, this area is considered not to require enhanced protection against drought and water scarcity. This national context does not encourage measures to anticipate drought induced by climate change. This particularly affects wet regions, where drought awareness is usually very low. In the context of the DROP project, it must be highlighted that IAV seeks solutions for drought adaptation (see Sect. 6.3.1), namely, by considering the implementation of drought forecasting and risk assessment tools. This represents a necessary initial step towards local preparedness for drought and local awareness (Richard 2013) of the possible impacts of climate change on drought. It is also a first modelling phase before initiating a common adaptation strategy resulting in adaptation measures.

IAV clearly has technical **resources and responsibilities** for the management of the Arzal dam and its reservoir. The objectives of the two pilot measures (Sect. 6.3.1) show the investment of resources by IAV towards the aim of managing a multi-use reservoir. However, information about the responsibility of each stakeholder remains unclear: the water users (especially farmers) are not always fully aware of the quantity and the quality of the water they are using. Local governments, who are responsible for various controls on water withdrawal, are not often aware of their duties. Concerning drought management, the regulation from the SDAGE indicates the priority to be given for various water uses in the event of drought. However, it does so at a very local level, while water policy (e.g. policy on pumping and pollution control) is clearly under the responsibility of the State services (national level). Thus, the extent of resources and responsibilities is positive, but the distribution of resources and responsibilities could be more clearly addressed.

6.4.2 Coherence: Agreement on the Priority to Give to Drinking Water

IAV has a central role in terms of cohesion among stakeholders within the Vilaine catchment. All interviewed actors agreed that having a single interlocutor, IAV, works well and facilitates dialogue. However, as far as the drought management issue is concerned, coherence between **levels and scales** has not been fully achieved: some actions are implemented coherently (e.g. interconnection between drinking water networks), but the integration of the climate change perspective lies mainly in the hands of the State services and remains inside the IAV through the coordination of the SAGE Vilaine.

Since the reform and decentralization of the role of the State in the 1990s, the State has had little decision-making power over water management issues. The decentralized services are managing water resources and making decisions within the catchment scale. They are under the authority of the Prefect, who implements the policy of the State under its regulatory and technical aspects. However, there is a lack of consistency between these government services in the Vilaine territory. Difficulties in coordinating interdepartmental relations have been cited during the interviews. At the same time, the **actors** in the Vilaine catchment are clearly used to working together, formally and informally. The Vilaine Committees (see Sect. 6.3.2) have been very successful at making people collaborate, and their success supports the coherence of actions relevant to water management at the catchment scale.

Regarding the common **perception of problems and goals ambitions**, the very positive consensus on water management can be assessed by the fact of general acceptance, i.e. the acceptance, by all stakeholders representing the various water uses, of the priority given to drinking water. The availability of water provided by the dam supports this priority. However, the perception of the risk of drought and the potential impacts on freshwater availability is almost absent for the majority of stakeholders. This can be explained by the more frequent flooding issues in the area, which was, in particular, observed during the period of interviews in connection with the floods that occurred in Brittany early in 2014.

As there is no comprehensive information about the quantity of water withdrawn by all water users, water managers are not really able to implement a truly coherent policy by soliciting coherent **strategy and instruments**. Moreover, the SAGE is a compromise between stakeholders, and some of them are more organized than others. As a result, some interests are taken into account more than others, especially those related to the drinking water supply. Furthermore, despite the work being conducted to integrate agricultural regulation in the WFD, agricultural policy is still declared by local stakeholders to be too thoroughly disconnected from the other instruments in the catchment due to the CAP (Common Agricultural Policy), which is decided at the European level and is not linked directly with water policy at the local level. A coherent implementation of strategy and instruments is, thus, hard to find for water management as a whole in the area and even harder for drought management.

The **responsibilities** of each actor with regard to the WFD are not obvious. Most actors think that the responsibility lies mainly in the hands of the State services as represented by the Water Agency. Most water users are not considering the link between their activities and water resource management. In fact, the LWC's **responsibility** is to be a decision-making authority, but it has neither the financial nor the technical means to implement decisions, so it has no **resources** to implement the plan of actions proposed. At the scale of the Arzal dam and its reservoir, the financial and technical aspects are provided by the IAV. At a more local level, we observe that some municipalities do not have the means to finance 20 % of the budget for the implementation of local projects related to water even if the subsidies, as those coming from the Water Agency, cover the remaining 80 %.

6.4.3 Flexibility: Limited by the Emergence of Multiple Structures Partly Compensated by the Number of Instruments

Most decisions about water management are made at the **scale** of the catchment and sub-catchment areas due to the decentralization of the role of the State. This situation can lead to a lack of consistency among the different services. In general, the State lacks the flexibility to better coordinate crosscutting issues. The current structures managed by the State work with a top-down approach. The two different speeds of operation and decision-making, the speed of the State and the speed of the LWC, represent a cogent reality for the LWC. The meetings held by the LWC are very pragmatic and achieve high goals in terms of water management, while State services struggle to advance equally rapidly and anticipate the coordination needs of stakeholders. At the catchment **level**, the LWC makes substantial efforts to integrate the different **levels and scales** of decision-making.

Actors and networks are all represented in the LWC, and there is real flexibility at the catchment level to allow the creation of committees devoted to special areas and topics. However, there is currently no committee for drought management and adaptation issues. Moreover, it appears that the priority given to drinking water is not flexible, as the quantity of water dedicated to this use needs legally to be delivered by IAV.

For the Arzal dam, there is little flexibility of **problem perceptions and goal ambitions** because of the importance given to the drinking water service. However, at the level of the catchment scale, the interconnection between the various drinking water networks allows some flexibility in water management even if it may not be sustainable on a long-term basis. Considering drought management because this is an issue that is not easily addressed by most of the actors, we can state the hypothesis that there is little flexibility in problem perceptions and goal ambitions related to drought.

At the local scale, there is a certain flexibility in adapting **strategy and instruments**. A concrete example is the change of the name of the Natura 2000 Commission initiated by the Mayors of the municipalities of the Vilaine catchment: they changed its name to "*Vivre les marais*" ("Living the marshes") as a way to gain more local support for the initiative. Additionally, "regional doctrine" for irrigation has been stated in the SAGE for the eastern areas of the Vilaine catchment, which are more vulnerable to low flows, under which the financing of water reservoirs is allowed only if there is no impact on the river flow during low-flow periods. This creation of new instruments seems, in fact, to be more flexible than the national policy. However, drought is not yet integrated into most of the existing instruments. Drought issues still rely on crisis management and not on anticipatory measures. However, given the impacts of low flows and drought periods on the various water uses of the Arzal reservoir, and considering the two pilot measures being developed by IAV, the flexibility of the multi-use reservoir is expected to be enhanced by decreasing the impact on water quality of the passage of the lock by boats and improving low-flow risk management. These measures would also imply a more favourable adjustment of the level of water in the dam for the wetlands during the dry periods.

The decentralization of **responsibilities** is currently an important issue in France due to a new governing body for the management of aquatic environments and flood prevention (GEMAPI), which will be transferred to local governments. Its creation is interpreted differently according to actors for the impact on

responsibilities and resources. For some, the GEMAPI will bring more coherence because it will allow good structures to be set up; for others, the municipalities would be forced to take on too many duties without sufficient financial means to implement the management measures. Considering the issue of drought adaptation, here again, the rigidity of responsibilities regarding crisis management does not allow flexibility in the implementation of drought adaptation measures.

6.4.4   Intensity: Awareness of Drought Issues Induced by Climate Change Is Low

The intensity of activation of several **levels and scales** is important in the Vilaine catchment for water management but not for the possible increase of drought frequency or intensity due to climate change. At the catchment scale, IAV has a strong impact in terms of preserving water supply and implementing some specific measures. However, IAV does not yet have the capacity to propose solutions for other water uses during drought periods. Furthermore, the low level of drought awareness among most of the actors is an obstacle that precludes the enhancement of the intensity with which levels and scales are recognized and implemented in drought adaptation strategies.

The intensity of one **actor**, IAV, is maximally strong in the lower Vilaine River area. It drives a **network** represented by all the representatives of water uses within the catchment and succeeds in making people work together throughout the area. This is mainly due to the coordination activity of the SAGE Vilaine, linking water users and water managers at the entire catchment scale surrounding the IAV. However, as the place of IAV remains too central, it can be interpreted also as indicating a certain vulnerability of the network.

Regarding **drought perceptions and goal ambitions**, we observe that awareness has gradually increased, although the anticipation of drought events is only effective in sensitive areas located on the eastern part of the catchment. Therefore, this awareness does not directly concern the Arzal dam and its reservoir. Only drought crisis management, based on critical observed low-flow conditions, is planned and integrated within the dam management rules driven by IAV. We can also emphasize that drought is seldom mentioned by stakeholders in relationship to climate change impacts. Up to now, climate change has neither been discussed within the LWC nor considered a threat to the sustainability of the production of the freshwater resource in the whole catchment. Only recently, at the end of the DROP project, has it received more attention in the management of freshwater in the Arzal dam.

In France, **strategy and instruments** for water crisis management are well defined at the State level. At the catchment scale, the SAGE regulation can provide additional guidance, such as the volumes of surface water bodies or groundwater to be allocated to each category of users. Regarding the Arzal dam, the abundance of water in the reservoir and the priority given to drinking water are sufficient to make it unnecessary to dedicate extreme effort to drought adaptation strategies. From a different perspective, several stakeholders around the Arzal dam mentioned during interviews that there is a need for groundwater monitoring to assess not only the state of the resource but also the level of consumption. In fact, there is actually a lack of groundwater surveys in the catchment area. There is neither an assessment of groundwater tables nor knowledge about the number of private wells and the associated withdrawals at critical periods. Thus, even if the issue of drought is receiving increasing consideration, the lack of effective instruments could impede the rapid development of a strategy of adaptation to droughts.

6.5   Overview and Visualization of the Results of the Analysis

6.5.1   The Priority Devoted to Drinking Water Production

In the Vilaine River area impacted by the management of the Arzal dam and its reservoir, one can say that the need to produce drinking water is paramount, and other activities related to water are placed at a second priority level. All users of water resources and local municipalities are aware that resource protection is crucial, but many of them do not want to bear the consequences of a restriction on water use in their territory—the "not in my backyard" syndrome. The need to produce drinking water is essential, but it also requires concessions. In the Arzal area, the only adjustable variable considered is the water level in the reservoir and, still, all stakeholders and managers agree on the priority to be given to the protection of water resources for producing drinking water. Trade-offs are only accepted when water scarcity becomes obvious, as in sensitive areas in the eastern sub-basins of the Vilaine River.

The observations described in the previous section lead to the general conclusion that the governance context for drought adaptation policies and measures for the case of the Arzal area of the Vilaine River can be considered as moderately positive. This general conclusion is obviously relative to the observations collected during the interviews that were carried out in 2013–2014 and may evolve with time. The following matrix seeks to represent the supportive, neutral and restrictive elements of the drought governance analysis provided (Fig. 6.7).

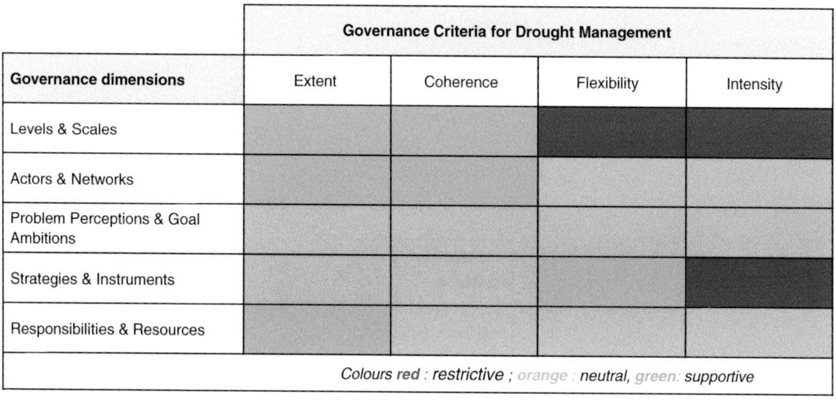

Governance dimensions	Governance Criteria for Drought Management			
	Extent	Coherence	Flexibility	Intensity
Levels & Scales				
Actors & Networks				
Problem Perceptions & Goal Ambitions				
Strategies & Instruments				
Responsibilities & Resources				
	Colours red : restrictive ; orange : neutral, green: supportive			

Fig. 6.7 Visualization of the main conclusions of the assessment of the drought governance context in the Arzal dam and reservoir, Vilaine catchment, France

6.5.2   The Interplay of Stakeholders and Their Motivations, Cognitions and Resources

The recognition of the needs for drought adaptation in France as a whole is strong at the national level, whereas it appears to be in only an extremely early phase in this northwestern part of France. The Governance Assessment Tool, based on Contextual Interaction Theory (see Chap. 3), enables a comparative analysis of key governance factors representing the freshwater use as the main water use. The governance context influences these processes through its impact on these actor characteristics: the drivers of processes are ultimately people, representing sometimes themselves, but often organizations or groups and themselves driven by their *motivations, cognitions* and *resources*.

In the Vilaine catchment, except for emergency measures, there is no global plan set up to manage drought vulnerabilities induced by climate change. The current situation of low drought risk perception, compared to a more significant flood risk perception, is explained by a lack of drought risk awareness, due to the absence of critical drought events in the past years in the region and the lack of a culture of drought forecasting and risk communication. However, it is expected that as drought perceptions are raised, drought adaptation measures can rapidly be designed and implemented by the efficient, existing water governance for freshwater in the basin, which is supported by a dense stakeholder network driven by IAV.

In the territory affected by the management of the Arzal dam, a key issue is that the acceptance of climate change (*cognitions*) as a reality or at least as a relevant issue for the stakeholders involved is very weak. This is identified as a major problem and a root cause for the low degree of openness towards adaptation (*motivation*). However, plain interests also play a role in this low motivation of

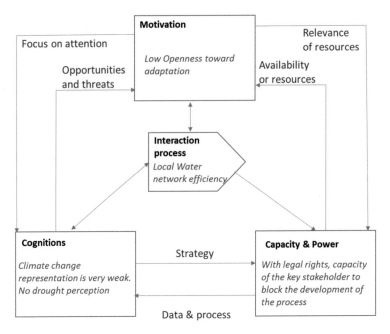

Fig. 6.8 Contextual interaction theory applied to the Arzal dam management in the Vilaine catchment, situation during interviews (2013–2014)

some of the users. With their legal rights (*resources*), they are also in the position to block the development of the process, at least until a higher level of awareness (*cognitions*) has been developed (Fig. 6.8).

However, after our diagnosis of the intensity of perceptions of local challenges for water use due to the impacts of climate changes, although the role of strong resources in blocking the development of adaptation measures could be considered highly negative, that is just a matter of interpretation. If the answer is "strong", we must recognize that the influence on the dependent variable that the score is very negative. This characteristic of the interaction process can also be highly supportive. Perhaps the story for Vilaine, as also formulated in words, is that all three main factors now have a negative influence on the progress of adaptation. However, when change occurs, most likely a change from *cognitions* (more awareness), to *motivation*, such change would have a positive influence on drought adaptation. Thus, the same strong *resource position* that is now capable of blocking progress would then become a productive position in the development of adaptation measures. As a consequence, relatively rapid change is not impossible and even possible. To represent this possibility, Fig. 6.9 describes the process induced by an improvement of local climate change awareness.

Finally, the last DROP meeting with practitioners, planned in the presence of French national observers (ONEMA, The French National Agency for water and aquatic environments) and held in June 2015, permitted an overview of the situation

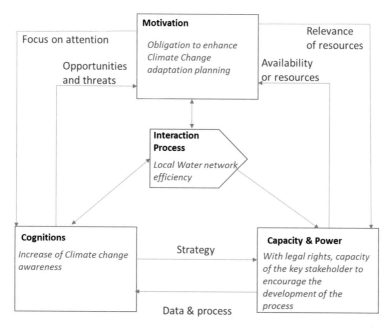

Fig. 6.9 Contextual interaction theory applied to the Arzal dam management in the Vilaine catchment resulting from climate change impacts awareness (scenario)

of drought awareness after one year. A significant change was noted in the consideration of the possible impacts of climate changes on drought frequency by IAV. It is not possible to affirm that this change was due to the work performed during DROP, but one can state that a supportive process facilitating the development of drought awareness is ongoing.

6.6   Conclusions and Case-Specific Recommendations

The governance team of the DROP project concluded that the strong network driven by IAV for drinking water management of the Arzal reservoir in the lower Vilaine River catchment and for the implementation of the SAGE management plan can also be used for drought management, even if this currently is not the case. If the focus is placed on setting up drought policy and drought adaptation plans, the structure of the system of the levels and actors in the whole Vilaine catchment will be available for its implementation. It is expected that it is the overall high availability of water as well as the common priority given to drinking water by these actors that currently forbid the implementation of measures for drought adaptation in the Arzal reservoir area except in those sub-basins where water scarcity is already obvious but where the use of water from the Arzal reservoir is not a concern.

Additionally, it must be noted that the recent reforms related to local governments introduced uncertainty in the actor network. In particular, this consideration is linked to the fact that the departments are not supposed to be involved in water issues in the near future. Additionally, there could be a change through the agency of the recent law on the modernization of public action (MAPAM), which would lead to a reaffirmation of everyone's skills and formal competences at the local level.

The matrix presented in Fig. 6.7 should be considered as a "photograph" of drought governance at the period of the interviews rather than an established state of drought governance for the Arzal dam and reservoir. Some recommendations have been proposed to enhance drought resilience through adjustments to current water governance in the Vilaine catchment. These recommendations are based on comparing the Vilaine governance context with previous knowledge from the Governance Team of the DROP project on other water management systems, including the ones investigated in the other regions studied in the project.

Taking into account the main supportive elements of the water governance in the Vilaine catchment, i.e. the ability to put together a strong network of actions to develop an integrated water management at the basin scale, the recommendations presented below are oriented towards a better incorporation of more "drought concerns" into the current governance scheme. They are also devoted to enhancing intensity and flexibility within the water governance framework. Five main recommendations have been formulated as follows:

6.6.1 Create a Task Force Dedicated to Climate Change Impacts on the Territory, Within the Existing Water Management Network, to Raise Awareness About Drought

In the case of the Vilaine site, as with other sites of the DROP project, there is a clear lack of sensitivity towards drought issues, most probably due to a highly favourable situation in terms of water availability in the area until the present time. An outreach effort is necessary to strengthen the awareness of water users and water managers of drought and adaptation measures that can be implemented in the area. Such an effort could first target a better understanding of the impacts of climate change in the specific territory of the Vilaine catchment relative to its own climate variability and vulnerability to drought and water scarcity. The efficient network of actors in water management, which is one of the strengths outlined in the case of the Vilaine catchment, could then be fruitfully mobilized around drought issues and around water scarcity more generally.

In this context, the LWCs driving the SAGE have a special role to play in mobilizing human and financial resources and interfacing with local actors. Geographically defined or topical committees can play a special role by tailoring the building of knowledge to specific challenging situations, e.g. estuaries, Natura 2000. Moreover, the creation of a task force dedicated to climate change impacts is recommended at the LWC level to keep this initial interest more closely focused on local needs and perceptions.

This sensitization of the actor network in water management towards climate change issue would enable increased awareness among all stakeholders regarding the potential impacts of climate change on water and water-related activities in the Vilaine catchment.

6.6.2   Enhance the Knowledge of the Water-Related Impacts of Climate Change in the Specific Vilaine Catchment

The task force mentioned in the preceding section would collect all the data related to the impacts of climate change on the territory of the Vilaine. These data should be used to identify and, if possible, quantify drought issues related to climate change and translation impacts in terms of water availability as well as their consequences for current and potential future water uses.

On the basis of this state-of-the-art data collection, the task force should also promote knowledge of climate change's impacts on the territory by undertaking specific studies targeting the main water uses of the catchment. Three main water sectors could be investigated: (i) agriculture, (ii) drinking water supply and (iii) tourism (including boating for the Arzal dam). Such a development would require a better knowledge about the interconnections between surface and underground water resources and could be supported by the monitoring of withdrawals, which has not yet been initiated. Moreover, it will be important to link flood- and drought-associated risks so that the solutions for flood prevention do not worsen drought situations and vice versa.

6.6.3   Develop a Strategic Foresight Analysis to Identify the Potential Types of Drought Situations in the Basin and the Means to Better Prepare Local Stakeholders to These Situations

Going a step further, enhanced knowledge about the impact of climate changes on the Vilaine catchment could be promoted by developing strategic foresight studies that would analyse future scenarios and help stakeholders to better anticipate the consequences of human activities on the ecological status of the basin and the quality of freshwater.

Within the SAGE instrument, an initial identification of sensitive areas related to low-flow water was conducted, especially for areas located at the eastern part of the

Vilaine catchment, which led to the development of specific rules. A comprehensive approach could be undertaken regarding the sensitivity of the area to the potential impact of drought related to climate change. This could be extended to the whole area of the catchment and contribute to establish a typology of geographical sectors and activities sensitive to climate change impacts. A more accurate assessment of the vulnerability of these sectors and territories, i.e. a precise evaluation of the consequences of drought and water scarcity for each type of activity and each location in space, would allow stakeholders to better anticipate the impact of a reduction of water resources on their activities.

6.6.4   Support the Development of Integrated Drought and Water Scarcity Management

Finally, a foresight analysis could lead to the identification of the most efficient measures to be implemented to take into account the potential climate change impacts on the Vilaine catchment into the economic activities of the water uses. This would support the development of integrated drought and water scarcity management, considering drought impacts on surface water as well as on groundwater and soil moisture.

To this end, a drought management and adaptation plan could be elaborated to address the impacts not only on surface water but also on soil and agricultural practices. In fact, the most recent droughts in the region were agricultural droughts. As far as water is concerned, a more comprehensive and effective policy aimed at reducing water consumption and withdrawals can be recommended. However, anticipatory drought management relies not only on the regulation of water level and withdrawals but also on techniques that, both, survey and help to keep moisture in the soil, which is of paramount importance for agricultural activities.

By identifying in advance drought adaptation measures to be implemented, anticipatory vulnerability assessments of surface water, groundwater and soils would complement the only measures of crisis management presently observed in the catchment regarding drought issues.

Drought adaptation plans could also be related to other planning documents, such as the one related directly to climate change, which usually relies upon other types of actors (i.e. SRCAE instrument cited in Sect. 6.2.2). Joint actions between the LWC and those actors already involved with climate change studies are highly recommended.

6.6.5   Sharing Low-Flow Forecasts with Reservoir Management Interested Parties

One of the aims of the DROP project in the Arzal dam site is the development of a tool for forecasting low flows and enhancing water management at the reservoir. Sharing information is essential to ensure that the management of the reservoir meets the standards of openness and transparency. It can also engage early cooperation within actors and postpone water shortage situations. Furthermore, numerical tools can help to corroborate decisions or to provide evidence of risks that may not have been foreseen, contributing to scientific arguments to address potential conflicts among stakeholders. As such, the implementation of a reservoir inflow forecasting and risk visualization tool may provide more flexibility to water management at the level of the Arzal dam.

References

Bressers H, de Boer C, Lordkipanidze M, Özerol G, Vinke-De Kruijf J, Furusho C, La Jeunesse I, Larrue C, Ramos MH, Kampa E, Stein U, Tröltzsch J, Vidaurre R, Browne A (2013a) Water governance assessment tool with an elaboration for drought resilience. DROP project, University of Twente, Enschede
Bressers H, de Boer C, Lordkipanidze M, Özerol G, Vinke-de Kruijf J, Farusho C, La Jeunesse I, Larrue C, Ramos M-H, Kampa E, Stein U, Tröltzsch J, Vidaurre R, Brown A (2013b) Water governance assessment tool. http://doc.utwente.nl/86879/1/Governance-Assessment-Tool-DROP-final-for-online.pdf. Accessed 14 Dec 2015
Bressers H, de Boer C, Lordkipanidze M, Özerol G, Vinke-De Kruijf J, Furusho C, La Jeunesse I, Larrue C, Ramos M-H, Kampa E, Stein U, Tröltzsch J, Vidaurre R, Browne A (2013c) Water Governance Assessment Tool. With an Elaboration for Drought Resilience. Report to the DROP project, CSTM University of Twente, Enschede
Brochet P (1977) La sécheresse de 1976 en France. Aspects climatologiques et consequences. Hydrol Sci Bull XXII 3(9):393–411
Corti T, Muccione V, Köllner-Heck P, Bresch D, Seneviratne SI (2009) Simulating past droughts and associated building damages in France. Hydrol Earth Syst Sci 13:1739–1747
European Drought Centre EDC (2013) European Drought Impact Report Inventory (EDII) and European Drought Reference (EDR). http://www.geo.uio.no/edc/droughtdb/. Accessed 22 Oct 2015

Le Roy Ladurie E, Rousseau D, Vasak A (2011) Les fluctuations du climat de l'an mil à aujourd'hui. Fayard, Paris, p 321

Mérillon Y, Chaperon P (1990) La sécheresse de 1989. La Houille Blanche N°5 (Août 1990), 325–340. doi:10.1051/lhb/1990025

Pirard P, Vandentorren S, Pascal M, Laaidi K, Le Tertre A, Cassadou S, Ledrans M (2005) Summary of the mortality impact assessment of the 2003 heat wave in France. Eurosurveillance Monthly Release 10(7):554. http://www.eurosurveillance.org/ViewArticle. aspx?ArticleId=554. Accessed 20 Oct 2015

Plan National d'Adaptation au Changement Climatique 2011–2015 (2011) ONERC, p 73

Poumadère M, Mays C, Le Mer S, Blong R (2005) The 2003 heat wave in France: dangerous climate change here and now. Risk Anal 25(6):1483–1494

Richard E (2013) L'action publique territoriale à l'épreuve de l'adaptation aux changements climatiques. Thèse de doctorat Université de Tours, p 520. http://www.theses.fr/ 2013TOUR1802/document. Accessed 16 Dec 2015

Robine JM, Cheung SL, Le Roy S, Van Oyen H, Herrmann FR (2007) Report on excess mortality in Europe during summer 2003. EU community action programme for public health, p 15

UNEP (2004) Impacts of summer 2003 heat wave in Europe. Environ Alert Bull (no2, United Nations Environment Programme Nairobi)

Vidal J-P, Soubeyroux J-M (2008) Impact du changement climatique en France sur la sécheresse et l'eau du sol. In: Magnan J-P, Cojean R, Cui YJ, Mestat P (eds) SEC 2008—International symposium drought and constructions, vol 1, Laboratoire Central des Ponts et Chaussées, Marne-la-Vallée, France, pp 25–31

Vidal J-P, Martin E, Franchisteguy L, Habets F, Soubeyroux J-M, Blanchard M, Baillon M (2010) Multilevel and multiscale drought reanalysis over France with the Safran-Isba-Modcou hydrometeorological suite. Hydrol Earth Syst Sci 14:459–478

Zaidman MD, Rees HG, Young AR (2002) Spatio-temporal development of streamflow droughts in north-west Europe. Hydrol Earth Syst Sci 5:733–751

Chapter 7
Flanders: Regional Organization of Water and Drought and Using Data as Driver for Change

Jenny Tröltzsch, Rodrigo Vidaurre, Hans Bressers, Alison Browne, Isabelle La Jeunesse, Maia Lordkipanidze, Willem Defloor, Willem Maetens and Kris Cauwenberghs

7.1 Introduction

This chapter presents a summary of the analysis results of the governance of drought-related issues in the Flanders region of Belgium. In the context of the Interreg IV-B project DROP, a team of researchers from four universities and knowledge institutes visited Flanders twice to perform interviews with authorities and stakeholders (October 2013 and May 2014). The visit was supported by colleagues at the Flemish Environment Agency (Vlaamse Milieumaatschappij, VMM). The exchange was held in the form of individual and group interviews and workshops with stakeholders including representatives from different institutions and sectors, e.g. from the drinking water company, national and local nature protection organizations, local farmers and local and national farmers organizations, the Flemish Environment Agency, different provinces, e.g. Province Vlaams-Brabant, and local municipalities, e.g. Kortemark Municipality. The analysis was guided by the drought-related Governance Assessment Tool (GAT) developed for the project. The GAT contains five governance dimensions (levels and scales, actors and networks, problem perceptions and goal ambitions, strategies and instruments, responsibilities and resources) and four governance

J. Tröltzsch (✉) · R. Vidaurre
Ecologic Institute, Pfalzburger Strasse 43/44, 10717 Berlin, Germany

H. Bressers · M. Lordkipanidze
University of Twente, PO Box 7658, 8903 JR Leeuwarden, The Netherlands

A. Browne
University of Manchester, Manchester M13 9PL, UK

I. La Jeunesse
University Francois Rabelais, Tours, France

W. Defloor · W. Maetens · K. Cauwenberghs
Vlaamse Milieumaatschappij, Aalst, Belgium

© The Author(s) 2016
H. Bressers et al. (eds.), *Governance for Drought Resilience*,
DOI 10.1007/978-3-319-29671-5_7

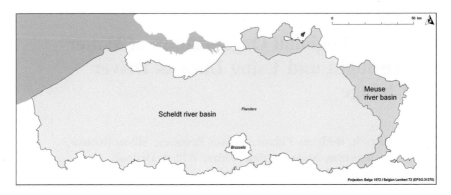

Fig. 7.1 Map of Flanders, with its territory's distribution over the two river basins Scheldt and Meuse (*Source* VMM)

criteria (extent, coherence, flexibility and intensity). The methodology is described in detail in Chap. 3.

The chapter presents the context of water management in Flanders, describes some measures which are already implemented related to drought management, explains the results of our analysis in terms of the Governance Assessment Toolkit and presents our possible conclusions and recommendations for improved drought governance.

According to the definition of the Water Framework Directive (WFD), the Belgian territory belongs to four international river basin districts (RBDs) which are shared with other member states and/or third countries. The RBDs of Meuse and Scheldt cover most of the Belgian territory, whereas the Rhine and Seine river basins cover much smaller parts in the south of Belgium.

Figure 7.1 shows that the Flemish region is located predominantly within the Scheldt river basin district. A comparatively small fraction of Flanders lies within the Meuse river basin district.

The Flemish region is divided over four river basins: Scheldt, Meuse, Ijzer and Polders of Bruges. Ijzer and Polders of Bruges are two comparatively small coastal catchments, added to the (International) WFD Scheldt River Basin District. The Flemish region covers 13.521 km^2 and has a population of 6.35 million inhabitants. The Scheldt (466 inh./km^2) and the Meuse (258 inh./km^2) river basin districts show a very high population density.

Contrary to the other chapters based on the DROP case studies, this chapter covers Flanders as an entire region. This choice was made due to the importance of federalism in Belgium and the more centralized approach compared to other chapters which are focusing on a specific catchment.

7.2 The Regional Organization of Drought Management: Flemish Water Management

7.2.1 Water Management in Flanders

Belgium is a federal state with responsibilities for water management at the regional and the federal level. The federal and regional competences are exclusive and equivalent, with no hierarchy between the standards issued by each. The Federal Government has amongst other things environmental responsibilities for coastal and territorial waters (from the lowest low-waterline). The Regions are responsible in their territory for environment and water policy (including technical regulations regarding drinking water quality, responsibility for the economic aspects of drinking water provision, land development, nature conservation and public works and transport. With the exception of the *Federal Plan on Coastal Waters*, water management plans are developed at the regional level, and therefore a mainly regional approach to river basin planning is used in Belgium. Coordination of water management planning occurs at the national and European level.

At the regional level in Flanders, three Flemish ministries are involved in integrated water policy: the Ministry of the Environment and Nature, the Ministry of Mobility and the Ministry of Spatial Planning. Many tasks related to integrated water management are assigned to the Flemish Environment Agency (VMM).

For the organization and planning of integrated water management, the Decree on Integrated Water Policy (of 2003) distinguishes three levels:

- The two International River Basin Districts (Scheldt and Meuse).
- The Flemish region with its four river basins (Scheldt, Ijzer, Polders of Bruges, Meuse—of which IJzer and Bruges Polder are added to the Scheldt).
- The 11 sub-basins.

This Decree is the juridical implementation of the WFD. However, it incorporates even more policy items of integrated water policy than those legally required by the WFD, prescribing more detailed planning on the level of sub-basins, as well as integrating quantitative aspects and the relation with spatial planning. It also contains the juridical implementation of the Floods Directive.

The responsibility for drawing up river basin management plans for the two Flemish parts of the international river basin districts Scheldt and Meuse lies with the Flemish government. The Coordination Committee on Integrated Water Policy (CIW; chaired by VMM) is designated as the competent authority for the implementation of the WFD, as well as for the Floods Directive. Among its responsibilities are the preparation of the River Basin Management Plans (RBMPs), including the sub-basin parts and groundwater specific parts, for the Flemish Region, reporting to the European Commission on WFD implementation, organizing the public consultation of the RBMPs, preparing the methodology and guidance for the development of the RBMPs and aligning the RBMPs with the Flemish Water Policy Note. The CIW consists of the executive management of the

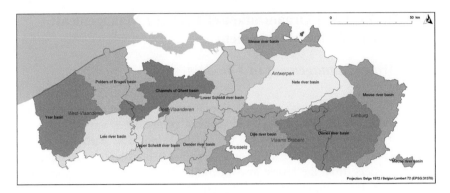

Fig. 7.2 Overview of sub-basins (11) in Flanders. In *red* are shown the limits of the Flanders' provinces (*Source* VMM)

administrative entities involved in water management. The CIW also oversees the functioning of the sub-basin structures, providing support and reviewing possible contradictions between binding provisions of the management plans at the different levels.

Figure 7.2 provides an overview of Flanders' eleven sub-basins, as well as the borders of the five Flanders provinces.

The international Meuse River Basin extends over five Member States of the European Union (France, Belgium, Netherlands, Germany and Luxembourg), and covers all three of the Belgian regions (Flanders, Wallonia and Brussels). The multilateral coordination in the international river basin district (IRBD) Meuse falls under the Meuse Treaty,[1] which regulates, among others, the international coordination of the implementation of the WFD in the IRBD Meuse. Further bilateral consultations are in place for the Netherlands and the Walloon Region.

In addition to these multilateral and bilateral consultations there is also intra-Belgian coordination for which the Coordination Committee for International Environmental Policy (CCIM) is used. This cooperation agreement is legally binding for the government after it was ratified by each government through law, decree or ordinance.

7.2.2 Evolution of Flanders' Water Policy

Starting in the early 1970s, Belgium started to progressively evolve towards federalism, becoming a fully federal state in 1993. In the field of water, the

---

[1]The Meuse Treaty was concluded in 2002 in Ghent between the governments of France, the federal state of Belgium, the Walloon Region, the Flemish Region, the Brussels Capital Region, the Netherlands, Germany and Luxembourg.

competencies for water quality aspects were transferred to the Regions in 1980, and
the same occurred in 1990 for the quantity aspects.

The Region of Flanders' initial approach to water management was the implementation of the 1971 ('federal') law on the protection of surface and groundwater
against pollution (Bressers and Kuks 2004). In 2003 the Integrated Water Decree
was passed in Parliament. The WFD is transposed with the coming into effect of the
Integrated Water Decree. River basin management planning is an important part of
the Decree.

The Integrated Water Decree applies to the water systems which are situated in
the Flemish Region. Water systems are defined as 'a coherent and functional entity
of surface water, groundwater, water beds and banks, including all living communities therein with all physical, chemical and biological processes thereof, and
the corresponding technical infrastructure' (CIW 2003). Thus, the scope of the
Flemish Decree covers surface waters and groundwater, as well as infrastructure
such as bridges, dikes, locks and dams. Furthermore, the Decree contains regulations on water quality management as well as on water quantity management. In
2010, the implementation of the European Flood Directive was integrated in the
Decree.

In accordance with the Decree, one minister in the Flemish Government is
appointed to be responsible for the coordination and organization of the integrated
water policy. He will be assisted by the CIW chaired by VMM. This
multi-disciplinary commission unites different levels of water management and
governance. It is responsible for the preparation, planning and the monitoring of
integrated water policy, and it is responsible for the implementation of the decisions
on integrated water policy of the Flemish government. The CIW also watches over
the uniform approach to the management of each basin.

The First Flemish Water Policy Note was prepared by CIW in 2005 and presents
the Flemish Vision on water policy (CIW 2005). The vision was updated in 2013
with the Second Flemish Water Policy Note (CIW 2013). This document holds the
goals of the Flemish government with regard to water management for the years
2014–2021. It names as its main goal the financing and implementation of integrated water management principles. It organizes the five current water management tasks in a framework of six guiding notions:

(1) Better protection and improvement of the quality of the water system:
    *Corresponding water management tasks: (1) the restoration of the good
    environmental status of the surface water and (2) the chemical water quality
    of groundwater resources require additional effort.*
(2) Sustainable management of water resources and ensuring sustainable water
    supply: *Corresponding water management task: (3) the water use requires
    further actions for improving water use quantity.*
(3) Integrated management of water scarcity and flooding: *Corresponding water
    management task: (4) the damage caused by water scarcity and flooding
    needs to be minimized further.*

(4) Further development of the vision to financing water management: *Corresponding water management task: (5) Big challenges will have to be coped with limited means.*
(5) Further stimulation of multifunctional use of water
(6) Working together on a strong and coordinated water management.

7.3 The Flemish Geo-hydrological Context: Using Data for Cooperation

7.3.1 Drought in the Context of Water Management in Flanders

In Flanders, pressure on water resources is high, which is amongst others due to the high population density. Water managers have historically paid much attention to guaranteeing water supply and water quality and mitigating the risk of flooding. Nevertheless, drought and water scarcity resulting from drought, are well known problems in specific sectors relying on a good water supply, such as agriculture.

In the past, Flanders, as most European regions, has experienced droughts in the years 1976, 1996, 2003, 2006 and 2011 [for further information see Chap. 6]. In recent years, droughts have had several consequences in Flanders. On some occasions water extraction from the Albertkanaal has been restricted. The 1996, 2006 and 2011 droughts were recognized as agricultural disasters and affected farmers were financially compensated. Temporary restrictions have been placed on the draft of ships on the Meuse river and in 2003 also on the use of the locks. Furthermore, Flanders experienced also other problems related to droughts; due to the wildfire in the nature reserve Kalmthoutse Heide in 2011, about 600 ha of heathland has been burned.

Drought policy in Flanders is based on the implementation of the European guidelines laid out in the WFD (EC 2000), resource efficiency policy (EC 2011), Blueprint to Safeguard Europe's Water Resources (EC 2012) and the Communication on drought and water scarcity (EC 2007). The management of water resources in Flanders, and hence the implementation of measures concerning drought and water scarcity, is delegated to a number of water managers with specific competences. The Division of Operational Water Management of the VMM is competent for the management of about 1,400 km of the larger unnavigable watercourses. As part of the DROP project, the VMM pilot is focussed on the improvement of drought risk management and the improvement of knowledge and data collection, which were listed as policy options in the European Commission's Communication on drought and water scarcity (EC 2007).

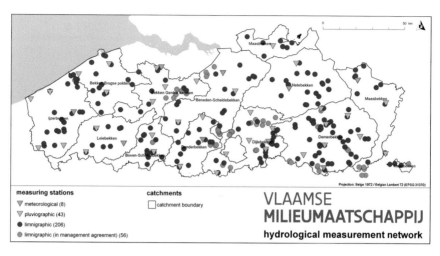

Fig. 7.3 Hydrological measurement network in Flanders (Belgium) operated by the VMM (*Source* VMM)

7.3.2 To Measure Is to Know: A Framework for Drought Monitoring and Modelling

VMM operates a dense hydrological measurement network to monitor precipitation, evapotranspiration, river stage and discharge in Flanders (Fig. 7.3). Real-time measurements from these stations are used for monitoring the water system, flood forecasting and operating the water infrastructure in Flanders.

In the DROP project, six water boards and water authorities implemented and tested innovative concrete measures focusing on specific drought and water scarcity problems as pilots. The aim of VMM's DROP pilot case was the development and use of indicators for the monitoring and reporting of the drought situation and the modelling of drought impacts using this measurement network. Also, further needs for the expansion of the network to make it better suited for drought applications such as drought monitoring, forecasting and the evaluation of drought measures, were outlined.

Droughts are complex phenomena, and are usually divided into four cascading levels: megadroughts; meteorological drought, resulting from a lack of rainfall; agricultural and ecological drought, resulting mainly from a lack of soil moisture; and hydrological drought, resulting from a lack of river flow. For the three levels: meteorological, agricultural and ecological drought, and hydrological drought, specific indicators were set up: meteorological drought is quantified by the Standardized Precipitation Index (SPI; McKee et al. 1993; Fig. 7.4) and Precipitation Deficit, agricultural and ecological drought by soil saturation, and hydrological drought by flow exceedance percentiles and the Standardized Streamflow Index

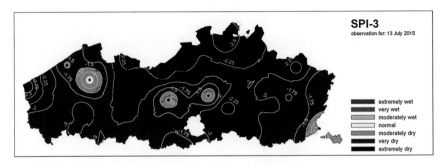

Fig. 7.4 Spatial distribution of the SPI-3 in Flanders on 13 July 2015 during a dry spell in summer (*Source* VMM)

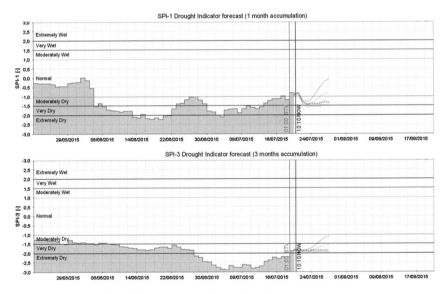

Fig. 7.5 Evolution of the SPI-1 (*upper graph*) and SPI-3 (*lower graph*) in the spring and summer of 2015 at the VMM pluviometric station at Boekhoute, with a 10-day deterministic forecast (*blue dotted line*) and probabilistic forecast (*grey lines* represent the 10th, 25th, 75th and 90th percentile of probabilistic forecasts) (*Source* VMM)

(SSI; Vicente-Serrano et al. 2011). Where possible, also a 10-day forecast for these indicators was developed to allow proactive management (Fig. 7.5).

To translate observations to impacts, modelling tools were developed. These models allow assessing the impact of past droughts (for a better understanding of drought impacts), present droughts (for operational drought management) as well as expected future droughts (in support of drought adaptation measures) on different aspects of the water system (e.g. soil moisture, streamflow) and on agricultural production (yield loss). In two pilot catchments spatially distributed SWAT (Soil and Water Assessment Tool, Arnold et al. 1998) water balance models and SWAP

(Soil-Water-Atmosphere-Plant) 1-dimensional soil moisture models were implemented. Through these models, the effect of soil and crop type, and drought severity and time of occurrence during the year on the eventual impacts of the drought can be highlighted. Further development of these models and expanding the modelled area can support the development of drought adaptation and mitigation policy in Flanders.

During the development of drought indicators and models for Flanders, a data and information gap concerning soil moisture, a key variable in drought monitoring and impact assessment, became apparent. Therefore, a soil moisture measuring campaign was set up, in which 15 traditional soil moisture probes and two innovative area-integrated soil moisture probes (COsmic-ray Soil Moisture Observing System; Zreda et al. 2012) were installed in the modelled catchments. In addition, the use of the Soil Water Index (SWI; COPERNICUS 2015) based on remotely sensed soil moisture data was tested in combination with ground measurements. The results of this study were used in the development of a Flemish soil moisture network, for which the probes will be relocated to locations distributed throughout Flanders to assure an optimal assessment of soil moisture conditions throughout the region in combination with satellite imagery.

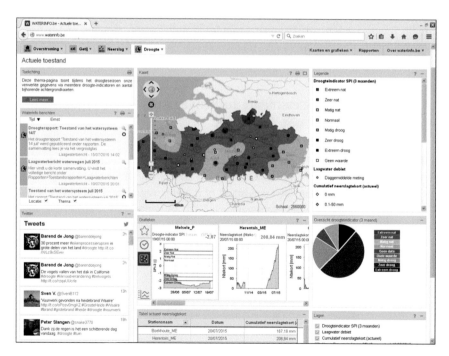

Fig. 7.6 www.waterinfo.be acts as a common portal for five Flemish water managers to report the state of the water system and focusses, depending on the situation, on floods, tides, precipitation or droughts (displayed here) (*Source* VMM)

7.3.3   Turning Data into Information and Cooperation

A further key step towards effective drought management is turning these data into useful information and getting the information to the different actors and stakeholders. Therefore, VMM publishes a monthly bulletin on the state of the water system, focusing on drought or flood risks as required. Since 2014, five Flemish water management services bundle their efforts to make real-time data and forecasts on the state of the water system available through the web portal www.waterinfo.be (Fig. 7.6). Drought is included as one of the four main themes of the portal. Information on the drought situation can be used by actors such as decision makers or water managers, or individual stakeholders, such as farmers, to evaluate the drought situation and take action accordingly.

It has been found that relying on the existing measuring network and framework of data management and reporting at VMM, contributes to a more effective drought status reporting. It also represents a technical step towards the development of an integrated water management strategy that addresses both high flows (floods) and low flows (hydrological droughts).

At the same time, the VMM pilot participated actively in the creation of a coordination platform for drought by bringing together different governmental agencies and organizations involved in water management and agriculture. This will further stimulate the cooperation between different actors and stakeholders on drought issues, such as the Flemish agricultural department, regional and national water managers, the provinces and municipalities.

7.4   Governance Assessment: Improvements in Drought Awareness but not There yet

In the following section, the four qualities of the GAT-tool presented in Chap. 3 are analysed for the Flanders region. Contrary to the other DROP case studies the GAT-tool was used here for the assessment of a whole region and not only a single catchment. Furthermore, the national level in Belgium has transferred most responsibilities to the regional levels: Wallonia, Brussels and Flanders. Because of the entry point at a relative high scale, one difficulty was to also cover small, individual activities on local level. Additionally, the relationships between different authorities and ministries were central in the discussions.

7.4.1   Extent

In terms of *extent*, the general water management system in Flanders is assessed as supportive with good involvement of EU level, Flemish region, provinces, communities and municipalities, as well as strong interactions between different actors and networks. The fact that the national (Belgian) level is receding to mainly a coordinating role is not considered to be a problem. The Flemish level is the key agent in generating initiatives and policy. However, there is a negative trend in which municipalities seem to be disengaging from water management responsibilities, for instance in their handing over responsibilities for "type 3" (small) watercourses to the provinces. This could mean that municipalities end up 'out of the loop' regarding, for example, water quality and nature protection, and thus may be unaware of opportunities for synergies related to water management, such as drought preparedness. This is problematic as they keep the responsibility for flood measures, and measures to dovetail drought prevention while addressing flood, may not be incorporated as they could. This is also relevant when considering that groundwater permits are given by towns and municipalities. This being said, interviewed stakeholders tended not to see this change as a problem.

Specifically in drought terms, *extent* is more limited, as initiatives are restricted to the higher levels, i.e. policy visions and planning initiatives for Flanders. The fact that recent European directives such as the Floods Directive, have a broader perspective which includes social and ecological criteria, has recently expanded the extent of actors and networks involved in water policy. The commitment to intersectoral involvement can be seen in the CIW, which includes leading officials of agriculture, nature and planning. However, involvement of stakeholders and the public in Flemish water planning is seen by many interviewees as closer to *pro forma* than to the aim of shaping environmental actions. Interviewees highlighted that they are not involved in the planning phase of policy, such as in the selection and prioritization of measures to be implemented, but only in the implementation phase itself. The wish to be involved earlier in the planning process was expressed clearly by several stakeholders. In addition, the lower level is not always actively participating in these processes, e.g. polders and wateringen, which are public authorities responsible for water resources management in the polder areas.

In general, drought is not yet an issue compared to the perception of flooding impacts for the region. The awareness of water scarcity and drought problems is very low for some stakeholders. Drought resilience, as a topic, is weakly developed in current constellations of problems and goals defined by stakeholders in the region. Various problem perspectives, e.g. from farmers, nature organizations, drinking water availability, are being taken into account in discussions, but there is as yet not much work on prioritization of drought and water scarcity as an issue.

In terms of instruments, different instruments are already implemented, such as groundwater taxes for business users (handpumps and households not included), groundwater permits, and restrictions for water extraction. Furthermore, important strategies integrate water scarcity and water demand, e.g. the 'Flemish vision on water policy' from 2005 to 2013 includes the sustainable use of water, and states that it is necessary to deal coherently with water shortage (CIW 2005, 2013). It has been agreed that the second River Basin Management Plans (due in 2015) should also include measures for water scarcity and droughts. In addition, the Environmental Policy Plan 2011–2015 'MINA-Plan 4' mentions as objective:

groundwater quantity and water use (Vlaamse overheid 2011). But a key missing instrument seems to be the agreement regarding flows over the border with France for the Scheldt river, and the lack of agreements regarding transboundary groundwater bodies. There is also significant potential for 'mainstreaming' drought into existing measures to make them serve multiple objectives. In particular, it would seem that flood prevention infrastructure (often small-scale dams) can be managed in a way that incorporates drought considerations. Otherwise, Flanders shows quite a significant number of instruments in place, particularly for groundwater management. However, with VMM which is the only actor developing measures, and with other stakeholders not having the issue of drought on their agendas, there does not seem to be a positive extent to the responsibilities and resources assigned. The level of involvement in planning is seen as insufficient by some stakeholders, who regret that they are not consulted in the planning phase of policy (measure selection and prioritization), but only when it comes to implementation questions.

7.4.2   Coherence

Coherence seems to be mildly supportive in the Flemish context, but this evaluation is mainly dependent on the direction of further activities. Whereas there are as yet not many actions that address droughts at the Flanders level, there seems to be a 'culture of coordination' within the authorities dealing with water topics, and other existing frameworks seem to be geared towards coherence. At the Belgian level, there are different coordination committees, but also institutional arrangements, such as the CIW, which meet frequently. The coordination between different authorities at the local level with district and province levels, seems to be both intense and fruitful, and planning moves up and down levels, according to the size of initiatives. The mentioned 'culture of coordination' shows that relevant actors work well together. There are good and productive relationships in place, with regular exchange. This is partly due to the long tradition of working together, that some actors have with the Flemish region. The last few decades have also seen great improvements in the dialogues with stakeholders traditionally less involved in decision-making (e.g. nature organizations). However, some smaller actors are more distant from the processes. At the lower levels, such as municipalities, the awareness seems to be quite moderate and partially a possible mismatch of objectives occurs.

Problem perspectives could be more coherent if knowledge-based approaches for droughts, such as the DROP pilot, are used. The ORBP process for floods showed that a solid common scientific basis can increase the coherence of perceptions. A need for a more integral vision on the water conditions of other sectors was already mentioned. Some incoherencies seem to exist between the drinking water companies (which provide lower water prices for large-volume consumers) and the environmental objectives of VMM.

Regarding instruments, the current existing instruments addressing droughts have not been developed within a strategic approach. The instruments were developed very independently for each of the different relevant areas. Because only a limited number of instruments relevant for drought purposes are in place, in general the instruments do not overlap. Smaller problems exist, e.g. between groundwater recharge and rainwater storage, which should be taken into consideration for new instruments. Over the past years, the topic of water scarcity and droughts has been integrated in several strategy papers and the Flemish vision on water. However, the future development of these issues could be helped by a more strategic approach. Further activities for a coherent approach in water management in general, can be seen in the synchronization of the planning period of sub-basin RBMPs (and basin RBMPs. Also the integration of drought measures in the RBMPs (from 2015) shows a tendency to a harmonized approach of developing and implementing drought measures with other water management measures.

The responsibilities for water scarcity and droughts are unclear or not assigned in many organizations and levels, including the Flemish regional level. Within in VMM, responsibilities are associated with different departments and therefore are very fragmented. This fragmentation leads to own individual discussions with the 'normally' involved institutions and levels. A coherent connection between the different discussions and approaches is not realized in the most involved organizations. Because the issue is not high on the agenda, there are as yet no competence conflicts between the different actors.

7.4.3 Flexibility

The analysis shows a moderate *flexibility* of the water governance system. Federal arrangements seem to point towards a good flexibility between different levels and scales, a flexibility that is built into the system: problems are dealt with 'at the relevant level', for example, in direct negotiation between Brussels and Flanders when it comes to a new waste water treatment plant. The possibility of issues moving up one level can be seen in cross-boundary management issues. However, on the topic of drought, the lead is currently firmly in the hands of VMM, and no clear possibilities of issues moving downwards, to a level closer to local, on-the-ground implementation of actions, were observed.

The high involvement of actors in the processes, with both formal and informal contacts, is positive. The formal coordination mechanisms are complemented by informal ones between different actors in water management. The fact that Flanders in a comparatively small region has all the different levels and actors ranging from a 'central government' to the local level would seem to help in the processes of VMM and others, reaching out to different actors. However, the flexibility in the kind of involvement and at what stage of the process could be increased, e.g. so that relevant actors can also be part of planning processes.

At the moment, goals are not at the stage of being discussed within policy processes, and therefore very far from being operational. Currently, the emphasis is still on building up a knowledge base. Willingness to find flexible solutions can be noticed, and water managers do meet and develop solutions, when necessary. This flexibility is also the result of not really having any formal mechanisms in place.

An element, conducive to flexibility, is the scenario-based planning approach. Furthermore, several instruments show aspects of flexibility. The short-term permits for groundwater in problem areas, with continuous search for surface water alternatives, show high degree of flexibility in the design and implementation of measures. Furthermore, the drinking water plans for increasing infrastructure interconnectivity show a tendency towards increased flexibility, as does the drinking water companies' recognizing the need for additional buffering capacity.

The scientific approach used for the development and implementation of flood measures, was driven by VMM. The stakeholders were presented with suitable measures, which had already been derived by VMM. In our interviews, some stakeholders criticized their involvement as being too late (e.g. not participating in measure selection) and mentioned that the acceptance of the measures is not always high. Therefore, the flexibility of the approach seems to be limited because the selection and adjustment of measures by stakeholders and implementers are not a main focus.

Because there is no assigned budget for droughts, the resources have to come from different related resources, such as nature and biodiversity; these include certain flexibility. Synergies occurring with a combination of drought and flooding are yet not taken into account.

7.4.4 Intensity

As in the other DROP case study areas in Northwest Europe, by far the relatively weakest point of the governance context for drought resilience policies and measures is its *intensity*.

The EU level, which has been fundamental in providing impulse to change in water management in Flanders, is only starting to develop relevant actions for water scarcity and droughts. There is some movement at the Flanders level, with droughts being incorporated as an issue into recent policy vision documents, but this has yet to be translated into actions. At the moment, the Flemish region carries the responsibility more or less single-handedly to bring in drought aspects into water planning, which makes activities dependent on this governance level.

VMM is keen to develop some of these actions and is the main driver behind initiatives to tackle drought—the DROP project being one of them. Although there is recognition of the importance and desirability of addressing droughts, at the moment no other actors really have this topic on their agenda or are driving it. There is only one actor driving change, and the establishment and implementation of strategies and instruments is very dependent of the Flemish regional level. At the moment, no strong drought-related strategy exists which could initiate strong drought-related activities. But in the last year, in several important strategic policy documents on Flanders level such as 'Flemish vision on water policy' and 'MINA-Plan 4', water scarcity and drought aspects were included.

Existing goals have to be improved. It will be necessary to go beyond the policy that is already in place (business-as-usual). Further goal ambitions should also be increased, but for this the awareness of the problem of water scarcity and droughts has to be improved so new and ambitious goals are agreed on with a huge number of relevant stakeholders. At the moment, many stakeholder groups have a quite low awareness of the problem and do not intend to support goals. It is currently only possible to implement local projects where local problems are clear. Even in these situations, gaining a long-term perspective versus just solving immediate problems, it is very problematic.

However, many stakeholder groups' awareness of the problem is quite low, and while they are in favour of initiatives addressing droughts, it is not considered as one of the issues on their agendas, which means that no energy and resources are earmarked to address the topic. Awareness of problems among farmers is growing, but they still want to use groundwater resources today and do not integrate the perception of future generations in their actions. The fact that most stakeholders do not seem to be involved, seems to be a hurdle for further actions, as much depends on individual coalitions between agencies and stakeholders. These coalitions have to be built up for every individual project implementation or also for policy processes. Furthermore, instruments for awareness-raising of different stakeholders are also not clearly defined.

7.4.5 Summary

The analysis of the drought governance qualities leads to the conclusion that the governance context for drought resilience policies and measures in Flanders can be regarded as at the moment 'intermediate'. Overall, both *extent* and *coherence* are between moderate and supportive for most governance dimensions, with the exception of 'responsibilities and resources', which are somewhat underdeveloped but are to be expected for a new issue. *Flexibility* is mostly moderate, with significant room for improvement, and *intensity* is as yet neutral to restrictive—again something to be expected for a new water management issue. In Fig. 7.7, our findings are visualized.

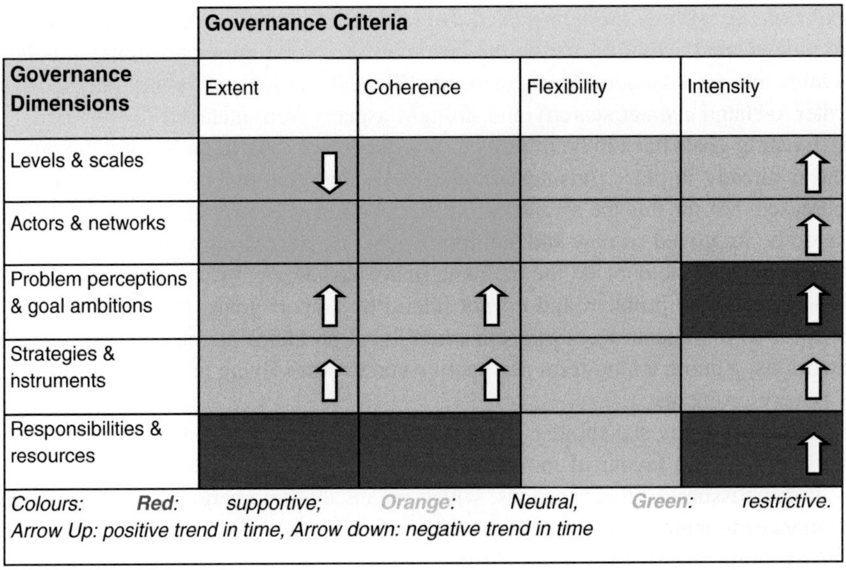

Governance Criteria				
Governance Dimensions	Extent	Coherence	Flexibility	Intensity
Levels & scales	⇓			⇑
Actors & networks				⇑
Problem perceptions & goal ambitions	⇑	⇑		⇑
Strategies & instruments	⇑	⇑		⇑
Responsibilities & resources				⇑

Colours: Red: supportive;   Orange: Neutral,   Green: restrictive.
Arrow Up: positive trend in time, Arrow down: negative trend in time

Fig. 7.7 Summary visualization—Governance context assessment for droughts in the Flanders region

7.5 Improving Drought Governance in Flanders: Conclusions and Recommendations

This section presents conclusions and a series of possible recommendations to improve, from a drought perspective, the water governance context in Flanders.

7.5.1 Overall Conclusions

The Flanders region researched in this chapter shows a limited awareness for the drought problem. By far, the most recognized problem in the region is flooding. The first actions are taken by VMM and some stakeholders, such as individual farmers, as far as agricultural droughts have been more frequent within the last decade. Thus, the VMM approach is mainly based on a scientific technocratic approach which produces modelling results to assess drought risks following the specificities of possible droughts (hydrological versus agricultural droughts or water scarcity). The estimated results are taken as basis for setting the policy agenda and development of activities. As described, Flanders takes forward a very scientific and centralized approach. The stakeholder involvement is already improving but is still limited. Measures are proposed and developed by VMM, stakeholders such as farmers, have difficulties to influence suitable solutions. Attention and regulations are missing at the federal level in Belgium which might be a reason why the Flemish region favours the centralized and less bottom-up approach. The Flanders level is the highest scale policy level at which the main objectives and strategies for water management are developed and at the same time the implementation of activities are covered.

The moderate supportive but partially restrictive governance setting can—if also in a limited manner—support the motivation, cognitions and resources of actors for implementing drought activities. Most actions are taken by VMM as Environmental Agency at the Flanders level which wants to push the topic but is also relying on a strong strategic vision still to be developed. Further bottom-up measures are taken by individual actors such as farmers who are installing water basins at their properties. So, motivation can be recognized partially in some actor groups. Resources are only available partially. Activities are relying on research money coming from external institutions such as the European Commission. The search for resources is taken on a case-by-case basis.

It can be summarized that Flanders is at an early stage of establishing drought resilience measures. First activities, e.g. by VMM, are starting and are initiating further motivation and processes. Water scarcity and drought is integrated in some general water management strategic documents.

7.5.2 Increasing Awareness for Droughts

Essential for further development of drought activities in Flanders seems to be an increase of awareness in different actor's groups, especially the groups which do not incorporate the risk in their operations. The most important audiences would be intermediaries and multipliers, such as farming associations or energy industry associations. The awareness should also be raised slowly so that in case of a drought event, at least a low media attention exists. Interviewees suggested that water pricing for actors, other than the broad public, is an important tool to increase awareness for resource scarcity. Awareness and additionally more involvement can also be increased if the flexibility to include stakeholders in the planning processes (especially selection, prioritization and calibration of measure) can be increased. Some interviewees suggested they see their involvement in current planning processes as occurring too late in the process. In other DROP case study areas (e.g. in the areas of the two Dutch pilots), authorities have had very positive experiences with strongly participatory approaches. The existing Flemish approach is focused on using models to make a scientific case for drought actions. It would seem valuable to expand VMM's strategy to include additional approaches, e.g. use of pilot measures, demonstration projects, showcasing actions, best-practice exchange schemes, and working more strongly in network-building and improving relationships with actors. Interviews showed significant potential to expand risk communication and information exchange to different economic sectors, with the objective to integrate drought risks and planning into private actors' activities. They could, for instance, take the form of dialogues on drought adaptation and could involve discussions on impacts and possible solutions.

7.5.3 Mainstreaming Drought Risks and Preparedness

With an integrated central vision for Flanders, drought management could be integrated in different environmental and related policies and could lead to a multi-objective planning of VMM but also a guidance for collaboration between Flemish Ministry of the Environment and Nature and other relevant ministries such as the Flemish agricultural ministry. Furthermore, mainstreaming into private actors' activities is missing in Flanders. Based on risk communication to private actors, the aim could be for a voluntary agreement or code of best practices for drought adaptation within and between sectors. The willingness of some actors seems to be possible. A requirement of drought contingency planning could initiate a process in the water-relevant economic sectors. Companies of these sectors could be required to prepare contingency plans for their operations during drought periods, thus increasing preparedness and reducing economic damage during drought periods.

7.5.4 Engagement with Other Public Actors

The relationship with other relevant authorities at the Flanders political level, e.g. with the Flemish Ministry of the Environment and Nature and the Flemish agricultural ministry could be improved by initiating further dialogues. A centralized strategic vision could also help for the cooperation with other ministries. All large Flemish rivers come from other regions. When addressing drought issues, the whole river basin should be taken into account, because Flanders is depending on activities of the upstream regions. Especially, discussions with the neighbours Germany and France should be focused. The discussion should take care of water quantity and quality. The engagement of municipalities seems to decrease; interviewees do not see it as a major problem due to the good connection between municipalities, the district and provincial level. Nevertheless, the impact of the disengagement cannot be foreseen at the moment; therefore we suggest observing the consequences this has on water management over time. Furthermore, the municipalities should be involved in the local showcases which are implemented on the ground and are also kept on board of the drought management process.

7.5.5   Evaluate the Importance of Data Availability Gaps and Prioritize Which to Address

Expanding the breadth of water management to include drought issues will require making available and collecting additional data, typically needed for an adequate evaluation of measures, their prioritization and monitoring their effectiveness. VMM could take on the task of analysing data availability and identify which data gaps should be addressed and in which priority. Data regarding volumes of water abstracted and used by surface water users (and sometimes even data regarding surface water rights) is often incomplete in Northwest Europe. This kind of data can be a key to identifying where the potential for drought adaptation lies, implementing measures that make use of that potential, and enabling private actors to incorporate drought risk into their actions.

References

Arnold, JG, Srinivasan R, Muttiah RS, Williams JR (1998) Large-area hydrologic modeling and assessment: Part I.Model development. J American Water Res Assoc 34(1):73–89.
Bressers H, Kuks S (eds) (2004) Integrated governance and water basin management. Conditions for regime change and sustainability. Kluwer Academic Publishers, Dordrecht
CIW (2003) Decreet betreffende het integraal waterbeleid http://codex.vlaanderen.be/Portals/Codex/documenten/1011715.html. Accessed 3 Aug 2015
CIW (2005) De eerste waterbeleidsnota. Erembodegem
CIW (2013) Tweede waterbeleidsnota—Inclusief waterbeheerkwesties. https://www.vlaanderen.be/nl/publicaties/detail/tweede-waterbeleidsnota-inclusief-waterbeheerkwesties. Accessed 3 Aug 2015
COPERNICUS (2015) COPERNICUS global land service—Soil water index. http://land.copernicus.eu/global. Accessed 28 July 2015
EC (2000) Directive 2000/60/EC of the European Parliament and of the Council of 23 October 2000 establishing a framework for Community action in the field of water policy, p 72
EC (2007) Communication from the Commission to the European Parliament and the Council—Addressing the challenge of water scarcity and droughts in the European Union. COM (2007) 414 final, European Commission, Brussels. http://eur-lex.europa.eu/legal-content/EN/TXT/PDF/?uri=CELEX:52007DC0414&from=EN. Accessed 14 Dec 2015
EC (2011) Communication from the Commission to the European Parliament the Council the European Economic and Social Committee and the Committee of the Regions A

resource-efficient Europe—Flagship initiative of the Europe 2020 Strategy. COM (2011) 21, European Commission, Brussels, p 17

EC (2012) Communication from the Commission to the European Parliament the Council the European Economic and Social Committee and the Committee of the Regions A Blueprint to Safeguard Europe's Water Resources. COM (2012) 673 final, European Commission, Brussels, p 24

McKee TB, Doeskin NJ, Kleist J (1993) The relationship of drought frequency and duration to time scales. In: Proceedings 8th conference on applied climatology, January 17–22, 1993. American Meteorological Society, Boston, Massachusetts, pp 179–118

Vicente-Serrano SM, López-Moreno JI, Beguería S, Lorenzo-Lacruz J, Azorin-Molina C, Morán-Tejeda E (2011) Accurate computation of a streamflow drought index. J Hydrol Eng 17 (2):318–332

Vlaamse overheid (2011) Milieubeleidsplan 2011-2015. Brussel. http://www.lne.be/themas/beleid/milieubeleidsplan/leeswijze/publicaties/Milieubeleidsplan2011-2015.pdf. Accessed 3 Aug 2015

Zreda M, Shuttleworth WJ, Zeng X, Zweck C, Desilets D, Franz T, Rosolem R (2012) COSMOS. The cosmic-ray soil moisture observing system. Hydrol Earth Syst Sci 16(11):4079–4099

Chapter 8
Drought Awareness Through Agricultural Policy: Multi-level Action in Salland, The Netherlands

Gül Özerol, Jenny Troeltzsch, Corinne Larrue, Maia Lordkipanidze, Alison L. Browne, Cheryl de Boer and Pieter Lems

G. Özerol (✉)
CSTM - Department of Technology and Governance for Sustainability,
University of Twente, PO Box 217, 7500 AE Enschede, The Netherlands
e-mail: g.ozerol@utwente.nl

J. Troeltzsch
Ecologic Institute, Pfalzburger Strasse 43/44, 10717 Berlin, Germany
e-mail: jenny.troeltzsch@ecologic.eu

C. Larrue
Université François Rabelais Tours, 37000 Tours, France
e-mail: corinne.larrue@univ-tours.fr

M. Lordkipanidze
CSTM - Department of Technology and Governance for Sustainability,
University of Twente, PO Box 7658, 8903 JR Leeuwarden, The Netherlands
e-mail: m.lordkipanidze@utwente.nl

A.L. Browne
Geography/Sustainable Consumption Institute, University of Manchester,
Arthur Lewis Building, Oxford Road, Manchester M139PL, UK
e-mail: alison.browne@manchester.ac.uk

C. de Boer
ITC (Faculty of Geo-Information Science and Earth Observation),
University of Twente, Drienerweg 99, 7522 ES Enschede, The Netherlands
e-mail: c.deboer@utwente.nl

P. Lems
Water Authority Drents Overijsselse Delta, PO Box 60,
8000 AB Zwolle, The Netherlands
e-mail: PieterLems@wdodelta.nl

© The Author(s) 2016
H. Bressers et al. (eds.), *Governance for Drought Resilience*,
DOI 10.1007/978-3-319-29671-5_8

8.1 Introduction

This chapter focuses on the Salland region of the Netherlands and presents our analysis regarding the role of governance context on the new irrigation policy of the Water Authority of Groot Salland (*Waterschap Groot Salland*—WGS). The irrigation policy was adopted in early 2013 by the five water authorities in the eastern Netherlands.[1] Given the drought conditions in this region, the policy is concerned with finding a balance between the use of groundwater and surface water by farmers and the water needs of vulnerable nature areas.

The outline of the chapter is as follows: In Sect. 8.2, an overview of the water management system in the Netherlands is provided. Section 8.3 presents the case study background, starting with the national policies and mechanisms that are related to the irrigation policy, continuing with the historical and political background of the irrigation policy. Then, in Sect. 8.4, a brief description of the water system of the Salland Region and the pilot measures that have been carried out in the Salland Region are described. In Sect. 8.5, findings from the application of the governance assessment tool on four qualities and five dimensions are discussed. Finally, in Sect. 8.6, the overall conclusion and a set of recommendations regarding the governance context of Groot Salland are presented, which can be useful for improving drought resilience in the region.

8.2 Water Management in the Netherlands

The Dutch water system is characterized by a complicated organizational structure that has been developed over the centuries of experience with collaborative and participatory approaches to water management. The current water management system involves various organizations that function at the local, regional and national levels. The major tasks related to water management, the responsible organizations and the financing mechanisms are summarized in Table 8.1.

The management of water resources and services is a public responsibility and comes under the public law. Four types of governmental organizations can be discerned regarding the management of water resources, namely central government, provinces, municipalities and water authorities. Water-related tasks that these organizations fulfil are financed by central funds from the government or from decentralized taxes. Additionally, (publicly owned) private companies manage the drinking water supplies at the regional level, which often implies serving for more than one provinces, and cover their costs under private law, while operating under the regulatory rule of the central government.

---

[1]These water authorities are Groot Salland, Regge en Dinkel, Velt en Vecht, Reest en Wieden and Rijn en IJssel. In January 2014, Regge en Dinkel and Velt en Vecht merged to form the water authority of "Vechtstromen" and in January 2016, Groot Salland and Reest en Wieden merged to form the water authority of "Drents Overijsselse Delta".

Table 8.1 Tasks, organizations and financing mechanisms of the Dutch water management system

Task	Organization	Financing
Flood protection, water quantity and water quality (main system)	Central government	General resources, pollution levy on national waters
Groundwater	Provinces	Regional tax
Flood protection, water quantity and water quality (regional)	Water authorities (public)	Regional tax
Wastewater treatment	Water authorities (public)	Regional tax
Drinking water supply	Drinking water companies (semi-public)	Price
Sewerage	Municipalities	Local tax

Source Dutch Water Authorities (2015)

Regarding legal provisions, the European Water Framework Directive and its daughter directives provide the overarching legal principles. The main legislation that incorporates the national needs is the "Water Act", which integrated various acts related to water and is in force since 22 December 2009. Several other legislations, such as the Water Supply Act and the Water Authorities Act, regulate the specificities of different sub-sectors of water.

Given the fact that the Netherlands is a delta country, the governance of water towards managing floods and protecting the society and the environment against flood damages, hence the "dry feet" policy, has been the ultimate priority of the Dutch water managers. Central government and water authorities are the two key actors that share the responsibility for flood protection. Despite the high priority on flood protection, the goals and priorities of the Dutch water management are quite diversified. This diversification can be attributed to the increasing pressure from the weather extremes associated with climate change as well as other relevant concerns such as the provision of sufficient drinking and irrigation water; protecting the quality of water resources; managing the level and quality of groundwater; and managing the complex web of waterways. The increasing attention for drought resilience, the core subject of this book, is a good example of this diversification.

8.3 From National Mechanisms to Regional Policies: Agricultural Needs and the Effects on Drought

8.3.1 National Policies and Mechanisms Related to Drought Adaptation

As outlined in the previous section, the water management system in the Netherlands involves various organizations from multiple policy sectors that operate at multiple governance levels. This multi-level situation also applies to the drought-related policies and mechanisms that are relevant for the Salland region.

The national Delta Programme has the ambition to solve the water management and security problems in the Netherlands. However, in the western regions of the country surface water is needed for flushing the water to prevent salt intrusion from the sea, while in the eastern regions, it is mainly used for irrigation. So far it has been a challenge to address such different and conflicting priorities of different regions.

On request of the Minister for Infrastructure and Environment, the national Advisory Committee on Water has issued an advice in March 2013, about the freshwater supply in the Netherlands, partly supporting the preparation of the Delta Decision Freshwater in 2015, which is part of the national Delta Programme. The committee regards it a public task to take care that there is and will be sufficient freshwater for all uses and nature, but this responsibility is bounded. When new big water users start in relatively vulnerable areas or when they demand water of a specific quality, it can be reasonable to demand also investments and co-responsibility from them. Furthermore, the country should prepare for situations in which the supply of freshwater is less self-evident. For the short run, it might be sufficient to optimize the water system. Next to that, innovations that lead to less water use and more water storage need to be furthered. Like with situations of acute flood risk, the committee also advises to have serious gaming exercises in which real decision makers and stakeholders practice with drought decision-making under stress to test for instance the efficacy of the "displacement chain" (*verdringingsreeks*).

The displacement chain is a policy guideline that stipulates which water uses gets priority when the freshwater supply cannot satisfy the demands of all uses. In this chain, the first priority is to prevent irreparable damage to the water system, the soil (for instance peat layers) or nature. Second in line are the drinking water and energy production utilities. Third are high value agricultural and industrial production processes and last are the interests of shipping, general agriculture, nature with resilience, industry, recreation and fishery. The displacement chain is not often used, since limiting some of the last priority uses, such as the irrigation of agricultural fields and gardens and car washing, has been generally sufficient.

Another relevant mechanism is the national coordination committee for water distribution (*Landelijke Coördinatiecommissie voor de Waterverdeling*, LCW). This committee consists of representatives from the ministry, including the public works agency, the Union of Water Authorities and the Interprovincial Consultation. They meet when the water level in the transnational rivers gets lower than certain values or when even without this being the case there are drought problems in several regions. Apart from proposing measures (in principle using the displacement chain, but also including fine-tuning of the water system where it can be regulated) they also issue "drought messages" to over 400 stakeholders whenever there are possible water shortage problems.

Apart from the Delta Programme, the new policies regarding drought and water scarcity need to be explained in the upcoming Water Management Plan 2015–2021, which also needs to respond to the European Water Framework Directive. The geographical level of this plan is the subbasin of Rhine-East. In the Dutch part of

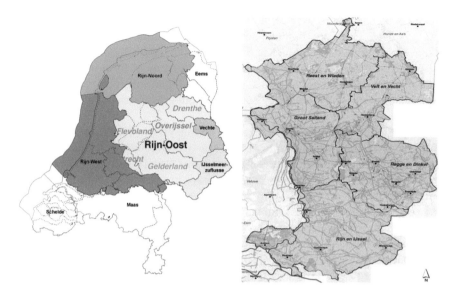

Fig. 8.1 The Dutch part of the Rhine-East Subbasin (*left*) and the five water authorities in Eastern Netherlands (*right*) (*Source* Anonymous 2015)

this subbasin, the management responsibility is shared among five water authorities, five provinces, Rijkswaterstaat (the national water agency) and about 95 municipalities (Anonymous 2015; see also Fig. 8.1).

As explained in Chap. 9 of this book, a manifesto was presented in June 2012 by the water authorities in the Rhine-East and the southern Netherlands, emphasizing the significant contribution of the higher parts of the Netherlands to agricultural production, and the importance of water management problems specific to these higher areas. The "ZON Declaration" (*Zoetwatervoorziening Oost-Nederland*: Freshwater supply East-Netherlands), which was signed in June 2014 as a follow up on this manifesto, brought together even a broader set of stakeholders including the provinces of Overijssel, Drenthe and Gelderland, the water authorities of the Rhine-East subbasin, municipalities, platforms for regional cooperation, nature organizations, agricultural organizations, drinking water companies and estate owners (Anonymus 2014). The ZON declaration formed a "political" statement to the national policy arena that the specific circumstances of the higher parts of the delta should not be underrepresented in terms of attention and funding.

8.3.2 Development of the Regional Irrigation Policy in the Eastern Netherlands

Agriculture has always been a key sector in the eastern Netherlands, especially after the Second World War, when feeding the increasing population emerged as a crucial issue and the memory of the famine during the last winter of the war was still very fresh. In the 1950s, investments were made to drain the agricultural lands and to enable irrigation for sufficient food production. As the farmers also observed the positive influence of irrigation on crop yields, many of them installed pumps and sprinkler systems on their fields. Thus, agricultural irrigation became a common practice, especially during dry summers. Both groundwater and surface water were used for agricultural irrigation.

Despite the positive influence of irrigation on crop yields, problems emerged in the 1980s. Drainage and groundwater extraction caused desiccation, which damaged both the agricultural areas as well as the natural areas that were sensitive to the changes in groundwater level. Improvement works were made to decrease the drainage of water from the land. By that time, however, irrigation was considered vital particularly for grass and crop production, since groundwater extraction reduced the watertable and in some areas the soil became too poor to produce food without irrigation.

In the 1990s and 2000s, environmental conservation became an important concern, partly due to the requirements of EU directives. Therefore, during drought periods, when the water sufficiency was threatened for drinking and industrial uses and for nature areas, water authorities were authorized to ban irrigation with groundwater. These bans caused problems for farmers, particularly those growing grass and corn.

As the water authorities started to see drought as a common problem in the region, they initiated discussions for a joint policy on the use of water in irrigation. They also had the additional objective of harmonizing their policies so that the farmers living in border areas of the water authorities would not be negatively affected of the different policies of different water authorities. It was, however, difficult to have all the water authorities on board. Some water authorities did not have problems during the major droughts, which were experienced in 2003 and 2010, so they did not want to spend time on developing an irrigation policy. Others had different priorities and they did not know where such a common policy would lead to. The water authority of Regge and Dinkel (WRD), which is now the Twente region of the water authority of Vechtstromen, was the only exception.

In 2010, WRD decided to have an irrigation policy, even if a regional policy would not be formulated. They concentrated on getting less irrigation from surface water and more from groundwater. When they spoke with the farmer organization (*Land- en Tuinbouw Organisatie*, LTO) they thought this is a solution that will have support with the farmers but this was not the case, especially because some farmers relied on surface water for irrigation. Thus, the LTO did not support the decision. Discussions lasted until the beginning of 2011, but no concrete outcome was reached. Then halfway 2011, the five water authorities decided to work together and initiated a project to formulate a policy that would address the protection of natural areas. They also involved the nature conservation organizations (NCOs), which were represented by their umbrella organization (*Natuur en Milieu Overijssel*, NMO).

In 2012 a consulting firm was hired to provide support on assessing the impact of irrigation on nature areas and defining a buffer zone around the nature conservation areas. A model was made on the influence of water extraction with respect to different soil types, extraction types and depths. The results of that model were used to decide on the size of groundwater-sensitive areas such as peats and swamps, which would be used as focal points, and the buffer zones, within which irrigation would be regulated. Several maps were made, including alternative buffer zones of 100, 150 and 200 m, and alternative sizes of nature areas with 50, 20, 15 and 10 ha. Each option implied different consequences in terms of area categorisation. The board members decided on the 200-m buffer zones, which would be defined around the nature areas of at least 20 ha. This was in line with a more general tone in Dutch government that emphasis should be put more on protecting larger nature areas and less on the smaller fragmented areas.

After consultations among the water authorities and with the provinces, LTO and NMO, the irrigation policy was issued in the spring of 2013. Although there were several issues to be worked out, the water authorities reached a decision, which was backed by their own boards. A major issue that was raised by the NMO was the lack of up-to-date information on water extractions in the buffer zones. In 2014, the water authorities made an inventory of the existing groundwater wells in the buffer zones to also identify whether and how much water is extracted without a notification, which is issued by the water authorities to the farmers for irrigating their land. Results from the inventory showed that the extractions in the current buffer zones are relatively small. Such an inventory was not conducted before the design of the irrigation policy, since the water authorities wanted to wait for the decision of the province of Overijssel as to whether water extractions in the buffer zones can or cannot be allowed. However, the province did not make this decision yet.

8.4 Too Wet and Too Dry: The Double Needs of the Salland Water System and Measures to Address This

8.4.1 Water System of the Salland Region

The jurisdiction of the WGS lies in the western Overijssel province, which is located in the north-east of the Netherlands. It constitutes a part of the Vecht/Zwarte Water catchment within the Rhine river basin. The WGS serves to a population of 360,000 inhabitants and numerous companies within a surface area of 120,000 ha. As shown in Fig. 8.2, the territory of WGS is divided into four districts, each of which has a district office and manages one or more wastewater treatment plants (WWTPs), in addition to managing a total of 4000 km of watercourse.

The area included within the WGS terrain has a 10-m slope, starting high in the Sallandse Heuvelrug and ending at the IJssel. The major canal in the area was originally dug for shipping. Some of the water draining from the Sallandse Heuvelrug, the higher elevation area, goes into this canal and some passes by underneath it. The system contains weirs and functions in an entirely regulated way by pumping water from the IJssel River upwards in the canal from Deventer onwards and then it trickles through the ground "downhill".

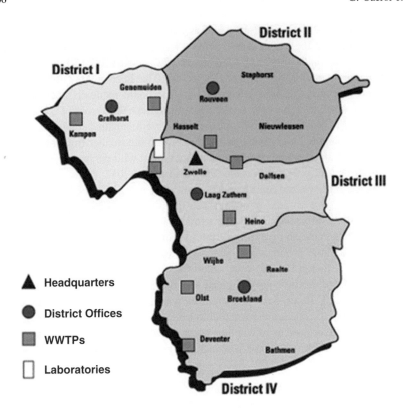

Fig. 8.2 Jurisdiction of the WGS (*Source* Waterschap Groot Salland 2014)

In accordance with the requirements of the Water Framework Directive, WGS has been working on a program to renew 37 watercourses before 2028. This program considers the bigger watercourses. WGS is also investigating opportunities for more extensive maintenance of watercourses to increase the robustness of the water system and hold the water as long as possible.

When the water level in the IJssel is too low for the pumps, they stop pumping (however they have lowered the pumps so this is required even less often than before). The amount of water that is permitted to be pumped is outlined in the agreement among the water authorities. Consideration is given to the levels of water required for fish in the various streams; however, there is no official requirement to do so.

Farmers have been accustomed to receiving the service of water being supplied for irrigation from the WGS. Since the WGS is responsible for determining how much water goes through the channels (due to the highly regulated nature of the system) when the channels go dry, they are ultimately responsible for having made that decision, except under extreme circumstances when higher-level regulations come into play. Thus, some farmers see it as the responsibility of the WGS to keep water in the channels. Until recently the WGS has had a general attitude of assuming such a supply-oriented role and doing their best to make sure that this was indeed provided. The development of the irrigation policy created an arena for additional stakeholders that address the competition between the wishes of farmers and other water users as well as a broader perspective to the optimal use of water by considering the issue of drought.

8.4.2 Pilot Measures Implemented Within the DROP Project

The catchment area of the pumping station Streukelerzijl, located in the north-eastern part of Salland, is prone to flooding due to an insufficient drainage system and to water shortage in periods of drought. The two pilot measures that are implemented in the Salland Region contribute to the protection of this catchment area of about 18,000 ha against both flooding and drought. This requires a double-acting system that is able to drain and supply enough water under, respectively, wet and dry weather conditions, and also a water system that responds quickly and effectively to changing weather circumstances. Another challenge of the project is to generate knowledge about how to enhance cooperation with all stakeholders involved in order to come up with new projects to prevent drought-related agricultural losses.

Within the first pilot measure, a large part of the catchment area has been disconnected to form a new catchment area in order to compensate groundwater extraction by the drinking water company and to secure the water supply for farmers in the catchment area. Two weirs and two pumping stations were built to discharge water to the Vecht River and to pump water from the Vecht River into the catchment area. The additional pumping station at the Vecht River, which will be built in the future outside the DROP project, will drain and discharge the new catchment area. Until this new pumping system fully works, a temporary water inlet is used to be able to supply the new catchment area with water. This temporary water inlet is located higher upstream than the future location of the new double-acting pumping station. The water drained from the new catchment area is temporarily transported via an already existing watercourse to the north-west.

The second pilot measure is an innovative system for managing the catchment area in a more efficient manner to prevent and reduce damage to agricultural production. It involves a remote-controlled steering mechanism that is linked to weather forecasts and to manage this management system it will be placed at the pumping stations. The WGS will start testing the steering mechanism at the two smaller double-acting structures as soon as possible.

8.5 Governance Assessment: After Acknowledgement of Drought Comes Integration of Drought

In this section we apply the Governance Assessment Tool (Bressers et al. 2013a, b) and present our findings regarding the four qualities of governance: Extent, coherence, flexibility and intensity.

8.5.1 Extent

The levels and scales dimension has a supportive extent, as all governance levels ranging from the local level to the EU level are relevant. However, the regional level is the most prominent, since the irrigation policy is binding for the five water authorities in the Rhine-East subbasin.

At the national level, the displacement chain, which is explained in Sect. 8.3, is implemented to balance the water supply and demand in cases of extreme water shortages. Interactions between different levels also occur for the implementation of relevant EU policies such as the Water Framework Directive and the Birds and Habitats Directives. The national level plays an enforcing role regarding the implementation of these policies. For instance, the Birds and Habitats Directives have requirements on the nature areas that are designated as Natura 2000 sites. The areas where the water level will be higher are identified in the provincial plan. Similarly, the measures that the provinces take for the nature sites are defined at the national level.

The extent is also supportive in terms of actors and networks. Many actors are involved in the decision-making processes, mainly including the province of Overijssel, the five water authorities, the LTO, the NMO and Vitens, the monopolistic drinking water company. Regarding groundwater management, the province and Vitens assume a role at the regional level. The province oversees the use and protection of groundwater by controlling the water authorities and issuing permits, whereas Vitens is interested in the influence on groundwater abstraction.

Regarding problem perceptions and goal ambitions, the extent is neutral. Many actors adopt a supply-oriented approach to water, implying that their major goal is providing the right amount and quality of water to all users. The focus on supply has been shifting since other interests became important in the past few decades. It was realized that the amount of freshwater is limited and climate change is exacerbating this issue. Drought is becoming a problem to tackle for the agricultural sector, although in many areas flood protection is still the major goal. Additionally, drought is a relatively new issue and hard to explain to the general public, although 1/3 of the country is similar to the land in the eastern Netherlands, which has dry sandy soils that are prone to drought. Farmers are affected by dry lands but do not consider this to be as important an issue in terms of "protecting their investments" that should be taken over by the water authorities through longer term and larger scale investments.

Several strategies and instruments are in place to deal with water scarcity and drought, indicating a high degree of extent. ZON is the first strategy that addresses the water problems of the East-Netherlands. It involves various measures such as using long-term climate data to demonstrate changes and engaging the relevant actors at multiple levels. The irrigation regulation is the major instrument regarding water use in irrigation. One aspect of the irrigation policy that is lacking in extent is that it only applies to new wells and not existing ones. For surface water management at the national level, the displacement chain is implemented. In terms of priorities, agriculture is the first sector that the water use is restricted through "irrigation bans", which are decided upon by the water authorities, whereas drinking water is the last. Permits and notifications are the main instruments to regulate water extractions (Ministerie van Verkeer en Waterstaat n.d.; Waterschap Groot Salland n.d.). The water authorities regularly monitor the groundwater levels and warn the farmers, in case of too much withdrawal.

The extent is neutral regarding the responsibilities and resources, since there is an imbalance between the large range of responsibilities, which have been assigned to various actors, and the often-limited level of financial and knowledge resources, which are decreasing for some stakeholders such as NCOs. Water authorities have the overall responsibility regarding drought adaptation. This is strongly reflected in the development of the irrigation policy. Both the water authorities and the provinces also have a responsibility for nature conservation areas, in particular the ones that are designated by the national government as Natura 2000 sites. Regarding groundwater use and protection, water authorities are responsible for shallow groundwater and the province is responsible for deep groundwater. The new irrigation regulation attributes some responsibility to farmers as well. They should notify the water authority when they will extract surface water and install a new well. While using water, they should also check the weirs and stop extracting surface water if the water does not flow over the weirs. Considering knowledge as a resource, there is a gap in the current understanding about the appropriate groundwater levels for agriculture and nature, and how they affect one another. It is also unknown how nature will adapt and how water extractions for irrigation will change under the new and developing conditions related to drought.

8.5.2 Coherence

Since the interdependence between different governance levels is recognized, a high degree of coherence is observed for levels and scales. However, coherence is low regarding the governance levels that are related to the environmental aspects of the irrigation policy. According to the national Nature Conservation Law (*Natuurbeschermingswet*), if someone implements a project in a Natura 2000 site, they have to prove that there is no environmental effect on the areas. In other water authorities, such as Brabant, where they have similar problems, farmers might have to take an additional permit from the province if their land is within a Natura 2000 site and have potential damage. However, the new irrigation policy does not incorporate the environmental impact assessment for Natura 2000 sites. The main reason for applying such a blanket approach is that the province had not decided yet how to deal with the extractions in the zones around the nature areas. There are 30–40 nature areas in the region, and the water authorities argue that it would take too long to consider the impact in each area. However, the EU policy would override the regional policy: If many farmers pump water to irrigate their fields and the groundwater level would drop, irrigation can be stopped according to Natura 2000 legislation.

A neutral degree of coherence is observed in terms of actors and networks. Since the numerous actors of water management have different interests and views, it is inevitable to have disagreements regarding how to allocate the limited water resources to different uses and services. For instance, water extractions threaten the groundwater-sensitive nature areas. As a result, water authorities, farmer organizations and drinking water companies take different sides than the NCOs. The "Agriculture on Sight" (*Landbouw op Peil*) project, which was started in 2011 by the water authorities and LTO, constitutes a positive example. This project involves individual farmers and aims to increase their awareness regarding the soil and water in their farm. According to an LTO representative, the project is changing the relationship between farmers and the water authority, who became more communicative and collaborative. The increasing cooperation requirements, for instance to implement the Natura 2000, is expected to create more initiatives similar to *Landbouw op Peil*.

Problem perceptions and goal ambitions are also neutral in terms of their coherence. The water authorities and the NMO have different perspectives on how the water system works and should work. On the one hand, the NMO sees the management of the water system leaning towards the interests of agriculture and not sufficiently protecting nature areas. On the other hand, the water authorities perceive the NMO as having strict positions on nature conservation and difficult to work with. This low coherence between economic and environmental goals seems to get higher as the irrigation policy takes into account the water needs of nature areas and limits irrigation around those areas. Furthermore, before the adoption of the irrigation policy, the situation regarding the extraction of irrigation water during drought, water shortage or calamities was not clear for farmers that live at the borders of water authorities and have land at more than one water authority. There are also differences in the perception of the urgency with which the issue of drought needs to be addressed, particularly between the water authorities and the NMO.

Strategies and instruments is the only restrictive dimension regarding coherence. The national Delta Programme was initiated mainly for solving the water management and security problems in the Netherlands. However, so far the different and conflicting priorities of the eastern and western regions of the country haven't been fully incorporated. Despite its apparent emphasis on freshwater supply, the ZON project addresses drought and potential measures. Permit is the main instrument regarding the regulation of water use; so its coherence with other instruments is relevant. The water authorities do not know which notifications were given in the protection zones. However, the irrigation policy will apply only to getting permits for new wells, not the existing wells. Even if it could be applied, the water authorities would have to pay to the farmers to move them out of the zone and render the permits unusable or the notifications invalid. Such a situation can also arise when Nature 2000 measures are implemented, since the province might have to pay to buy out the land around the protection zones.

The coherence of responsibilities and resources is neutral. Each actor has some responsibility regarding certain elements of the water system. However, it is unclear whether an actor holds the responsibility to connect all the knowledge regarding different aspects and has an overview of the water situation and the water balance. This can be explained by the very nature of water management in the Netherlands, which attributes more value to involving stakeholders from all levels and sectors, who have

8 Drought Awareness Through Agricultural Policy: Multi-level …

their own particular views about the system, and less value to assigning responsibility to a stakeholder for having a complete overview. The positions regarding the responsibility of water authorities in providing water to farmers differs between the water authorities that have sloping areas (WRD—farmers do not expect that WRD provides surface water for their fields) and those have mostly flat areas (WGS and Veld en Vecht—farmers do expect that water is provided). This situation makes the farmers in flat areas more advantaged than those in sloping areas, who have to incur extra costs to pump surface water to their fields in case they are not allowed to use the groundwater. These differences are not dealt among different water authorities.

8.5.3 Flexibility

A highly flexible governance setting is observed regarding the possibilities for upscaling and downscaling the policy issues between different levels. This results from the collaborative and participatory environment in the sector, which involves actors from different levels to elaborate on problems and solutions at multiple levels without imposing a hierarchy, unless there is a law that regulates otherwise. The development process of the irrigation policy constitutes a typical example of upscaling where the WRD scaled up the irrigation issue from the local level to the regional level, although it was not a major issue in WGS. Another example is the ZON, which downscales the national freshwater supply problem of the Netherlands to address the regional context, i.e. high sandy soil conditions.

The flexibility regarding actors and networks is assessed as neutral. Through designing an irrigation policy, the five water authorities were successful in developing a common regulation at the regional level. This process was completed without following a formal procedure and therefore it can be seen as an indication of a high degree of flexibility. However, the LTO and NMO were not able to effectively participate. There is little evidence that alternative participatory mechanisms (for instance surveys, polls, public hearings, focus groups, etc.) were applied for enabling the participation of individual or local actors in developing the policy. This aspect indicates a low degree of flexibility. Respondents from several stakeholders addressed the increasing emphasis on integrating multiple sectors, and thereby involving multiple stakeholders in projects such as *Landbouw op Peil.*

Problem perceptions and goal ambitions are also assessed as neutral. The way that the irrigation policy was formulated raises several concerns, in particular for the NMO, who sees the policy as incomplete and requiring changes according to the updated data on the existing permits and water withdrawals. Nevertheless, the water authorities have a strong trust in the policy-making process and are prepared to incorporate the influence of existing wells and groundwater withdrawal levels into the parameters of the policy. The same degree of flexibility applies for the concerns of the province on the incorporation of requirements by the national and EU environmental policy. In case the nature conservation policy changes and influences the irrigation policy, the water authorities are ready to amend the irrigation policy accordingly. The board members of WGS are generally associated with different groups or parties. There may be barriers to change when they go against the interests of the various organizations.

Strategies and instruments are also assessed as neutral in flexibility. As discussed earlier, the water authorities and the NMO have different opinions on the timing of the irrigation policy. The NMO defended a policy that would integrate all concerns and knowledge from the beginning, where the water authorities opted for starting the execution of the policy and adjusting later, if necessary. The water authorities took this decision based on the argument that the 200-m zone can be changed in the coming years if it proves to be too large or too small for some nature areas. The water authorities are also open to changes that can result from the implementation of other policies such as the Natura 2000. Water authorities and the provinces have the possibility to address the necessary changes in the five-year Water Management Plan.

Finally, the flexibility of responsibilities and resources is also assessed as neutral. The line between the practical responsibilities of the water authorities and those of the province is not always as sharp as it is on paper. The national Water Act, enacted in 2009, states the actors that are responsible for the different aspects of water management and draws the boundaries of responsibilities. However, for specific cases in the field, it is sometimes unclear whether the province and/or the water authority is responsible. Regarding water use at the farm level, the new irrigation regulation enables a flexible distribution of responsibilities. Instead of general irrigation bans imposed by the water authorities, the farmers are able to control water use according to the flow of water through the weirs. However, as mentioned above, this might cause discussions between the water authorities and the farmers, in case the farmers decide to take water when the water flows from the weir. In contrast to the very high degree of flexibility in terms of responsibilities, no such flexibility is observed regarding financial freedom. Provinces and water authorities have to deal with a situation where there is less financial room to play than in the past. If they want to do big investments, the provinces and water authorities have to get backing from their provincial parliament and general board, respectively. Similarly the NMO is dependent on the limited funding from the province, and needs to allocate it according to the priorities of the province.

8.5.4 Intensity

The irrigation policy receives attention and efforts from many levels, giving it a highly intense character in terms of levels. At the provincial level, the importance of groundwater is acknowledged for both the economy (the agricultural sector) and for nature protection (the areas that are dependent on groundwater). Additionally, the EU level creates pressure on the water authorities to work together. According to the Water Framework Directive, all these water authorities are in the Rhine-East and they need to collaborate on water planning and management. The history of good relations among the water authorities creates additional impetus regarding the enforcement of the irrigation policy. The regional approach is also reflected in the ZON declaration, which states that all parties in the East of the Netherlands will be responsible to optimise the availability and use of freshwater and to make efforts for decreasing the sensitivity of the region to extreme weather conditions.

A neutral degree of intensity is observed regarding the actors and networks. Drought is a priority issue for the five water authorities in order to cooperate at the regional level, so they communicate to find points of cooperation. Other actors, however, assign different levels of priority for several reasons. NMO cannot allocate time to projects on water since it has very limited financial and human resources. As a result, NMO cannot be proactive in terms of participation, but rather responds to the requests of the water authorities. The province has recently changed its approach to water projects. Since they have ongoing tasks and want to interfere in different policies in several sectors, they integrated water into the spatial planning sector. While they previously had a water team of 20 people, since 2012 they have a diverse team of 70 members. Finally, the LTO makes efforts to get involved in projects, although they lack the technical knowledge.

The problem perceptions and goal ambitions is the only restrictive dimension regarding intensity. The dilemma between putting the water away to prevent flood and letting it in the system to prevent drought is felt commonly by the practitioners. Furthermore discharging the excessive water is crucial, since getting the water out is more difficult than getting it in. Farmers also have higher priority for floods than droughts; they are worried when their fields are wet or when the groundwater is high, not when the land is dry. Nevertheless, drought constitutes a threat for agriculture due to the need for irrigation. Therefore, measures are taken to prevent farmers from facing water shortages. Furthermore, there is a national agreement that water from the rivers can be transferred from other regions. However, the water authorities differ in terms of the practicality and costs of transferring water. Another relevant issue is monitoring of water withdrawals. Until recently, there was no up-to-date data on the existing pumps, and thus no accurate information on where and how much water is being pumped. The provinces and water authorities use a national database to register the water users and update it once a year. But farmers usually do not comply with that rule and the water authorities do not regularly monitor the compliance, either. Water authorities need these data to manage the water flows and to monitor groundwater use. In case of water shortages, formally every farmer has to stop irrigation; otherwise, they will be misusing the water. This creates a kind of guarantee that not too much water will be withdrawn.

A neutral degree of intensity is observed regarding strategies and instruments. The intensity of strategies is increasing in terms of data management on water use. The database that the water authorities have is not reliable at the farm level, since the farmers did not provide the exact level of extraction; neither did the water authorities measure it. In order to update the database and to identify the unregistered wells, the water authorities made a survey with the farmers that have a land in the buffer zones or next to nature areas. On the other hand, a low level of intensity is observed regarding the irrigation policy, since it applies only to new wells. Although the water authorities considered banning the existing wells in the protection zones, they realized that this would be impossible, since some of the existing wells were not registered. If the existing wells would be banned, the water authorities might have to pay the cost of a new well and not irrigating the fields, as well as facing the resistance of many farmers that would want to continue using their wells. According to NMO, leaving all the old permits gives a "window dressing" character to the irrigation policy. They think that the policy is made for situations that might not so easily emerge, since the farmers are unlikely to invest in new irrigation systems due to installation and fuel costs.

The intensity of responsibilities and resources is also assessed as neutral. An increasing level of efforts is made towards the adoption of drought measures. The major indication of this increase is the ZON, which has been developed under the national Delta Programme. Within the scope of ZON, the local and regional stakeholders and the Delta Commission will contribute a significant amount of funding for implementing measures that will address both droughts and floods. The design of the ZON involves both using long-term climate data to demonstrate changes in the water system, and engaging the relevant actors at the local, regional and national level, who in the end agreed to devote resources for the realization of measures. The five water authorities have already allocated significant time to formulating the irrigation policy, and are committed to invest in implementing a metering system to monitor water use at the field level. This situation indicates a high degree of intensity. However, currently they do not have the resources to monitor or enforce the amount of water that is being taken by farmers (valid for WGS, unknown for the others). Furthermore, the influence of NMO is decreasing due to lower funds from the government and the province, while the water authorities expect NMO's inputs. In terms of NCO representation, two critical questions are raised: *Who pays for the voice of nature conservation? What if some actors do not want any more the voice of nature conservation?* NMO organizes the NCOs into the planning process and represent them during the meetings with the water authorities. This implies that if the water authorities involve the NMO, they can reach to all nature organizations.

8.5.5 Overview of the Assessment Results

Our observations demonstrate that the Salland Region has a neutral governance context regarding its drought resilience policies and measures. As visualized in Fig. 8.3, the context involves mostly neutral aspects, with five supportive elements and two restrictive elements.

Considering the five dimensions of governance, the most supportive one is "levels and scales", whereas the coherence of "strategies and instruments" and the

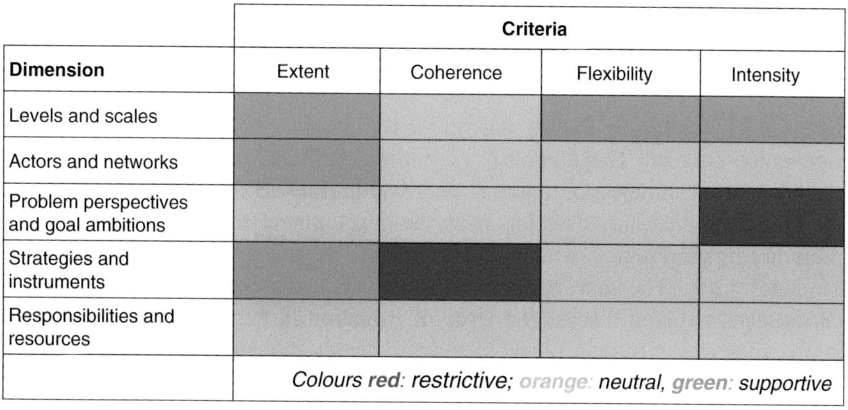

	Criteria			
Dimension	Extent	Coherence	Flexibility	Intensity
Levels and scales				
Actors and networks				
Problem perspectives and goal ambitions				
Strategies and instruments				
Responsibilities and resources				
	*Colours **red**: restrictive; orange: neutral, **green**: supportive*			

Fig. 8.3 Visualization of governance assessment conclusions

intensity of "problem perspectives and goal ambitions" are the only restrictive contextual factors. Regarding the assessment criteria, extent is the most supportive criterion, since three of the governance dimensions are assessed as supportive, while a neutral level is observed for two dimensions. Coherence is identified as the relatively weaker criterion compared to other three criteria, since it is assessed as neutral on four dimensions and as restrictive on one dimension.

Regarding coherence, the restrictiveness is mainly attributed to the fact that drought measures are not integrated into the existing water use, management and governance systems, partly due to the long-term competition that exists among different water user sectors (agriculture versus industry versus nature) and among different regions (east versus west). Nevertheless, the collaborative and trust-based atmosphere, which is developed through different projects and initiatives, is seen as a solid basis to reach coherent problem and system perspectives as well as collaborative and participatory mechanisms. The increasing understanding on the risks of drought for all water users creates a collaborative environment for all the stakeholders.

Regarding the restrictiveness of the intensity, the historically grounded concern on "too much water" and thus the dominance of managing the flood risk plays a major role. This historical context makes it difficult to diversify the priorities towards combating with "too little water". The water authorities invest time and money in improving their monitoring and enforcement systems towards better managing the system. However, actors such as NMO and LTO put relatively lower effort in such initiatives, due to a lack of financial resources and technical knowledge, respectively.

8.6   Conclusions and Recommendations for Salland: Seeking More Horizontal Integration and Awareness

Our overall conclusion is that the Salland Region is characterized by a vivid governance context. The emphasis of national and EU policies on river basin management encourages the regional water authorities to coordinate their actions. One of the regional initiatives has been the development and implementation of a common irrigation policy, which aims to balance the use of water by farmers close to natural areas. The investigation on the governance context revealed that all stakeholders involved discuss the issue of irrigation in terms of (a lack of) water supply, an approach culturally and historically firmly rooted in this region. As such, the focus of the policy shifted towards zoning, a solution that aims to reserve scarce water for nature, during periods of drought. However, the governance assessment also revealed some evidence that stakeholders on the regional level learn to treat the issue of drought as a phenomenon in itself through participating in a regional initiative, which aims to preserve and increase the freshwater reserves in the region, whereby stakeholders are willing to address drought as an issue in itself that influences the vulnerability and adaptability of their activities.

8.6.1   Influence of the Governance Context on Actor Characteristics

Looking from the actor characteristics perspective of the Contextual Interaction Theory, the relatively neutral governance context of the Salland region does not have significant positive or negative influences on the motivation, cognitions and resources of the actors that are involved in the implementation of drought adaptation. Water authorities have a thorough understanding of the drought problem as well as adaptation requirements. They support this by allocating resources and trying to mobilize all the actors. Furthermore, high degree of flexibility and intensity in terms of levels and scales enables the rescaling of the irrigation issue. However, the low degree of intensity regarding the problem perception leads to a situation that most of the actors, particularly the farmers, put much higher priority on flood protection than on drought adaptation.

Three major recommendations can be made towards improving drought resilience in the Salland Region.

8.6.2 Develop an Integrated Understanding and Approach to Managing Drought

In areas where freshwater resources are crucial both for agricultural production and for the protection of nature areas, the impacts of drought, such as low water levels and soil moistures, can be detrimental. Various policies and initiatives at multiple governance levels will have implications regarding the design and implementation of the measures for preventing and alleviating such impacts. The EU policies, such as Natura 2000, and regional initiatives, such as the ZON agreement and the irrigation policy, are at their infancy regarding the incorporation of drought adaptation and alleviation measures. The water authorities in the east and south of the Netherlands are recommended to use this opportunity for putting forward the specific context of the region in terms of drought vulnerability and intensifying their efforts for making sure that drought-related measures are sufficiently elaborated in these policies.

Another aspect that could benefit from an integrated approach is the treatment of flood and drought as separate policy issues. Despite the historic role of floods in Dutch water management, there is an emerging emphasis on the "double-goal" of managing flood and drought together. The pilot project that has been implemented by the WGS is a typical example of such an approach. Such integrated measures can be intensified when introducing other measures such as renovations in the water system and mechanisms for monitoring and evaluating the indicators on water availability and consumption. A final recommendation regarding monitoring and evaluation is the upscaling of monitoring mechanisms, for instance through creating system-level knowledge on the water budget to monitor sectorial water use (which are mainly agriculture and environment in the Salland Region) and define actions that can be taken by different actors at different levels. The complicated actor network of water management in the Netherlands makes it a big challenge to hold a single actor responsible for integrating all the knowledge regarding different aspects and for having an overview of the overall water resource situation and the water balance. However, as the pressure from drought impacts increase the competition among the water user sectors, development of such comprehensive monitoring mechanisms could be inevitable in the near future.

8.6.3 Raise Farmers' Drought Awareness Towards Creating Ownership and Drought-Sensitive Water Use

As in many other regions, farmers in the Salland Region can be key actors for reaching both economic and environmental goals. In this regard, the communication of drought-related information, particularly the drought-related risks, would be crucial. Information sharing tools that both deploy the technical knowledge and take into account the local knowledge and needs of farmers can be developed and made accessible to the farmers by also considering the legal requirements of creating and sharing such data. For instance, providing regular information to the farmers about the hydrological situation in their plots could directly increase their awareness about the drought conditions. It is also important to establish clear rules as to when and why farmers are not allowed to withdraw groundwater and/or surface water. For instance, decreasing groundwater levels is a local phenomenon: If the groundwater level drops in a field, it goes back to normal in a few weeks when it rains. With the new irrigation policy, farmers are not allowed to pump groundwater near a nature area, as this will negatively affect the groundwater level in that area. Farmers can easily understand and agree with such rules when the reasoning behind them and their relevance is communicated. Establishment of such rules would also indirectly contribute to another governance issue, namely the balancing of supply management with demand management, given that the current functioning of the water system is dominated by a supply-oriented approach. As the impacts of climate change are likely to put pressure on the availability and accessibility of freshwater resources, the management of the water demanded by farmers would become a major concern regarding the sustainable use of water in irrigation. Effective implementation of measures, such as the monitoring of notifications for groundwater extraction and the metering of water withdrawals at the field level, could contribute to the management of farmers' water demand.

8.6.4   Enable the Active Involvement of Non-governmental Organizations Towards Creating Shared Responsibilities

Non-governmental Organizations (NGOs), such as environmental NGOs and farmer organizations have positive intentions for improving the current situation, yet they lack the mechanisms and resources for representing their interests at higher decision-making levels. For instance, the LTO lacks the technical capacity to contribute to the debates on climate change in general, and drought in particular. Similarly, the NMO represents all the local NCOs in Overijssel, but its limited capacity in terms of financial and human resources leads to underrepresentation at the regional level. Their involvement is further threatened by the cuts made in the funds allocated for directly participating in relevant projects or initiatives. Active involvement of environmental NGOs and farmer organizations can broaden the perspectives for understanding drought and create more willingness to share risks.

Despite the expected benefits of increasing the involvement of NGOs, it is also acknowledged that many questions regarding division of risks and responsibilities would need to be addressed by changing nature of the involvement of these actors. The improvement of information sharing and communication mechanisms among the actors would be recommended for facilitating a fair and clear distribution of responsibilities. Given that the water governance system is open to designing new participatory initiatives, the ZON declaration can be instrumental for redesigning the role of NGOs in drought adaptation. The ZON declaration refers to the co-responsibility of all relevant stakeholders, while in its current form it is currently too broad to elaborate on how to share the responsibility among different stakeholders. During the process of stipulating the details and implementation mechanisms of the ZON declaration, it would be advisable to define mutually agreeable and feasible mechanisms to assign fair and clear responsibilities to all the involved stakeholders, with a particular emphasis on dissociating the level of responsibilities from the level of financial contribution.

Consequently, the main conclusions that are drawn from the Salland case and the recommendations that are made based on those conclusions pinpoint the significant role that multi-level actions play in drought adaptation.

References

Anonymous (2014) Intentieverklaring voor het Uitwerken van een Uitvoeringsprogramma Zoetwatervoorziening Hoge Zandgronden. http://www.helpdeskwaternl/onderwerpen/wetgeving-beleid/kaderrichtlijn-water/uitvoering-nationaal/rijn-oost/klimaat-droogte/@38658/intentieverklaring. Accessed 12 Nov 2015
Anonymous (2015) Geografisch gebied en waterplannen - Van Zevenaar tot Emmen: het deelstroomgebied. http://www.helpdeskwater.nl/onderwerpen/wetgeving-beleid/kaderrichtlijn-water/uitvoering-nationaal/rijn-oost/geografisch-gebied/#Waterbeheerenwaterplannen. Accessed 18 Aug 2015
Bressers H, de Boer C, Lordkipanidze M, Özerol G, Vinke-de Kruijf J, Farusho C, La Jeunesse I, Larrue C, Ramos M-H, Kampa E, Stein U, Tröltzsch J, Vidaurre R, Brown A (2013a) Water governance assessment tool. http://doc.utwente.nl/86879/1/Governance-Assessment-Tool-DROP-final-for-online.pdf. Accessed 14 Dec 2015
Bressers H, de Boer C, Lordkipanidze M, Özerol G, Vinke-De Kruijf J, Furusho C, La Jeunesse I, Larrue C, Ramos M-H, Kampa E, Stein U, Tröltzsch J, Vidaurre R, Browne A (2013b) Water governance assessment tool. With an elaboration for drought resilience. Report to the DROP project, CSTM University of Twente, Enschede
Dutch Water Authorities (2015) Water governance—the Dutch water authority model. The Hague, The Netherlands
Waterschap Groot Salland (2014) Beheersgebied. www.wgs.nl/groot-salland/structuur/beheersgebied. Accessed 3 Feb 2014

Chapter 9
The Fragmentation-Coherence Paradox in Twente

Hans Bressers, Koen Bleumink, Nanny Bressers, Alison Browne, Corinne Larrue, Susan Lijzenga, Maia Lordkipanidze, Gül Özerol and Ulf Stein

9.1 Introduction

In this chapter, we will concentrate on the Dutch water authority of Vechtstromen, more specifically the region of Twente part of Vechtstromen. The Twente region has some 135,000 ha and about 630,000 inhabitants. Though most of the Netherlands is flat and the highly artificial system of waterways often enables to let water in from outside each region, a substantial part of the Twente region does not have this option and is thus fully dependent on rainwater and groundwater. Apart from the wetland nature areas, especially the northeast of the region is for this reason relatively vulnerable for water scarcity and droughts.

In this paper, we will first explain backgrounds of the national drought governance in the next section. In Sect. 9.3, we will discuss the regional geo-hydrological context, drought policy focus and the measures taken, mostly in the framework of the DROP project. Thereafter, we will analyse and assess the supportive quality of the governance context for the implementation of these measures in Sect. 9.4. Section 9.5 concludes with a number of case-specific recommendations to reduce the restrictions and make optimal use of the strengths of the governance context.

H. Bressers (✉) · M. Lordkipanidze · G. Özerol
University of Twente, PO Box 217, 7500 AE Enschede, The Netherlands
e-mail: hans.bressers@utwente.nl

K. Bleumink · N. Bressers · S. Lijzenga
Waterauthority of Vechtstromen, PO Box 5006, 7600GA Almelo, The Netherlands

A. Browne
University of Manchester, Manchester, United Kingdom

C. Larrue
University Rabelais de Tours, Tours, France

U. Stein
Ecologic, Berlin, Germany

© The Author(s) 2016
H. Bressers et al. (eds.), *Governance for Drought Resilience*,
DOI 10.1007/978-3-319-29671-5_9

9.2 Dutch Drought Policy and the Needs of the "High and Sandy" Eastern Netherlands

In this section, we will describe a number of issues regarding drought governance in the Netherlands. Hereby, we will concentrate on what is relevant for the kind of drought and water scarcity problems that are typical for the part of the country that is dependent of rain water and ground water (like most of Twente) and the kind of preparatory drought resilience measures that are part of the DROP pilot projects in Twente. Thus, we will not discuss the possible water shortages to flush polder water ways in the west of the country to avoid salinization. Neither will we discuss the Dutch policies on the prioritization of water needs in times of urgent water scarcity and the management of irrigation. This subject has already been discussed in the previous chapter on the Dutch Salland region.

During the 1980s, the major environmental policy themes were mentioned in Dutch government white papers. This was the first time that drought was recognized in Dutch policy as a major issue, labelled as the theme of "desiccation". The emphasis was completely on the decrease in vitality of inland wetlands. The recognition of the problem at a national policy level did, however, not result in effective measures that solved or even stopped the gradual worsening of the problem. Furthermore, the more extreme weather conditions that are related to climate change are further increasing the vulnerability. Nevertheless, until recently almost all attention went to the risks of floods, not so much the risk of droughts and water scarcity. Perhaps not strange in a country where 55 % of the land is in principle flood prone. It is only recently that the already ongoing damages caused by drought receive more widespread attention.

In the white paper that started the reassessment of Dutch water management, the report by the Delta Committee of 2008 (Deltacommissie 2008: Samen werken met water, p. 71), there was just half a page of attention for the problematic of the "higher sand grounds". The committee plead for increasing use efficiency and more buffering of (rain) water and points to two investment programmes to enable this. While in the beginning the water authorities in the east and south of the country felt that their problematic did not receive sufficient attention, they cooperated to develop a "Deltaplan for the higher sand grounds" and organized a major conference with hundreds of people in 2012 issuing a manifesto "Water op de hoogte" (Water at the high level). Thereafter, attention for drought issues increased.

It was in June 2012 that the water authorities in Rhine East, together with their colleagues in the south of the Netherlands organized a big symposium Hoog en Droog ("High and Dry") where some six hundred people from involved governments, consultancies, business and NGO's participated, including the Delta Commissioner that steers the Delta Programming process. Here, a manifesto was presented in which attention for the special water management problematic of the higher parts of the delta was asked and further efforts and collaboration of all water authorities, provinces, municipalities and other societal organizations involved were announced. This impressive meeting had a real impact on the Delta Programming process and thereafter drought resilience was firmly positioned on the agenda. Until then "water shortage" was to a large extent seen in relation to the huge need for freshwater in the polder areas in the west of the Netherlands, which regularly need to be flushed to prevent intrusion of salinity. The manifesto a/o. stated that 45 % of Dutch agricultural value is in fact produced on the higher grounds.

In response to an invited advice from the national Advisory Committee on Water on freshwater supply, by 2013 the responsible Minister of Infrastructure and Environment emphasized that the system that supplies and distributes the water from source to user extends from the estuaries of big rivers all the way back to the capillaries of the regional water system. The steering of the fresh water supply should take place at all levels and scales: from the cross-boundary international river catchments up and to the local scale of individual users of stakeholder organizations. The main question is not at which level steering needs to take place, but which responsibilities are at which level and whether between those levels there is good collaboration.

In the Netherlands the implementation of drought policy is thus seen as a matter of needed cooperation between various organizations, both public and private, at various levels. This is in fact not just the case with drought policies but with other water policies as well. Perhaps it is typically Dutch to interpret the necessary coherence as a matter of cooperation rather than coordination by a powerful central actor (compare OECD 2014). However, the complexity and dynamics of the water system itself make a governance context that facilitates good cooperation by all stakeholders with their various interests very valuable. Also there is a need for productive boundary spanning between an inspiring long-term vision and short-term opportunities to realize parts of it.

While the Netherlands has been recalibrating its water policies in an enormous multi-stakeholder exercise called the Dutch Delta Programme, "fresh water supply" has become one of the main issues (sub-programmes). Droughts and water scarcity issues are not the same problem but they are actually highly related, while the water scarcity issues typically become most urgent in periods of drought. In the partial Delta programming on "Fresh water supply", the present and future policies are developed and implementation guided.

In the Delta Decision 2015 some attention is given to the problematic of the "High Sandy Grounds", the areas in the east and south of the Netherlands that are often not able to receive water from the main water system, and thus depend on rainfall and rain fed small rivers and creeks for fresh water supply. Drinking water companies, food industries, other industries and farmers use often deep and shallow groundwater for their production processes. It is recognized that these areas (including Twente) suffer from droughts for dozens of years and that climate change can worsen these problems further, causing dry creeks and damages to human uses and nature. The preferred strategy to combat these developments consists of the following guidelines:

1. Keep the water longer in both the ground and the surface waters. No efforts will be made to enable major water transport to these areas. In the short run, the

focus is on increasing the groundwater buffer and the moisture buffer at plant root level.

2. Saving water by more efficient water use. In the short run, by educating water users. In the medium and long-term periods of drought are unavoidable and consequently major water users should take measures themselves to avoid this from causing major damage.

3. Develop for the medium and long term some modest possibilities for extra water transport (a.o. to Twente via the Twente Canal where a brand new lock enables better to keep the level up).

All in all these guidelines show that the emphasis is on measures in the water system to increase buffer capacity both in the creeks and in the ground and on making water use more efficient. There is an own responsibility for users to decrease the potential damage from droughts.

Already before, in the document "Kansrijke strategieën voor zoet water" (Promising strategies for fresh water), September 2013, for the east of the country, including the province of Overijssel, the following measures are mentioned for the short run: smart "locking" (regulating the water levels), restructuring of the regional water system, making creek valleys wetter, increase groundwater storage and buffer water in larger nature areas. While drought resilience measures are often having spatial consequences and often deal with agricultural land and while they often take the form of renaturation of the water system, land use planning, agricultural policies (also from the EU) and nature, landscape and tourism policy sectors are very relevant.

Another relevant development is the national Administrative Agreement Water in which the state government, the provinces, the water authorities and the municipalities have stipulated the division of their tasks in water management and the way to integrate them and also agreed on the principles of cost sharing. In this framework, also the agreement has been made that the provinces and water boards will elaborate for the complete rural area the desired ground and surface water levels (GGOR). This specification can later be used as a justification for taking further measures. In the Delta Programme part on freshwater supply, emphasis is placed on the instrument of "specified level of provision". This instrument does not necessarily entail that desired water levels are specified in a quantitative way, but is seen as sets of information on the water system and its likely developments, combined with agreements between all relevant stakeholders about measures to be taken and about how to deal with remaining drought and water scarcity risks. The ambition is to have such stipulations and agreements for all areas by 2021.

Apart from the Delta programme the new policies regarding drought and water scarcity also need to be explained in the upcoming Watermanagement Plan 2015–2021, that also needs to respond to the European Water Framework Directive. The geographical level of this plan is the subbasin of Rhine East, containing six Dutch water board areas, including that of the merged Vechtstromen, but also consisting of the relevant provinces, municipalities, drinking water company and state water agency representatives. Of special relevance for the East of the Netherlands, including the Twente region, is that on June 27, 2014 (just before the second GT site visits to Groot Salland and Vechtstromen) the so-called ZON Declaration (ZON is an acronym representing: Freshwater supply East Netherlands) was signed by the provinces of Overijssel, Drenthe and Gelderland, the water authority of the Rhine East region, municipalities, platforms for regional cooperation, nature organizations, agricultural organizations, drinking water companies and estate owners. In this declaration they all acknowledge their co-responsibility for "an optimal availability of freshwater, a responsible use thereof, and the task to make their water system more resilient for extreme weathers". This co-responsibility also involves the preparedness to contribute financially to the cost of the programme. One of the actions involved is the specification of the water service level that users can expect in order to "clarify the role and responsibilities of the governments and the risks and behavioural options for the water users". The Declaration is seen as a major step forwards in the collaboration between all partners involved, but also as a "political" statement to the national political arena that the specific circumstances of the higher parts of the delta should not be underrepresented in terms of attention and funding. The declaration was successful in obtaining some funding from the national Delta programme, though less than hoped for.

9.3 Dry Creeks and Measures Taken in the Twente Region

9.3.1 Twente's Drought and Water Scarcity Situation

The following overview will concentrate on the regional level of the Twente region of Vechtstromen and more specifically on the relatively higher area where all of the many local pilot projects are located, the northeast and east of Twente. These projects and other activities and measures in the same area taken together are almost completely representing the drought resilience policy of the water authority in the Twente region (Bressers et al. 2015).

In the Twente region, all the drought and water scarcity problems as expressed in the "higher sandy grounds" initiative are present. Ninety percent of the small creeks are running dry in summer (Fig. 9.1) and when nothing is done this will probably increase with climate change (in the first eight months of 2013 rainfall in the Netherlands was 37 % less than "normal", seven of the eight months had shortages). It causes for instance complains from both nature organizations and farmers. Flora and fauna in the creeks die, and surrounding nature is suffering. Yields can fail and in cities algae bloom can occur. Extraordinary dry years were for instance 2003, 2006, 2009, 2010 and 2013. But already before, in the period 1994–1996, for three consecutive years irrigation bans had to be announced.

Desiccation has a great impact on the aquatic nature. Lower groundwater levels reduce the river discharge with the impact particularly seen in spring and summer

Fig. 9.1 Creeks in the Twente region that run dry each summer (*red*) or have less discharge than desired (*yellow*) (*Source* Website Regge and Dinkel)

periods, when precipitation is low and river levels are maintained by groundwater. Fish die and also the risk of algae bloom increases. Not only in streams and their valleys, but also on high ground in the catchment area nature will deteriorate with further desiccation in drought periods that are expected to become more frequent because of climate change. In the whole catchment area, substantial impacts are expected. Partly these problems have been aggravated by earlier measures some decades ago of the Twente water authority itself. Fighting water problems in wet periods by "improving" the drainage capacity of the water system increases vulnerability for droughts. The challenge is now to create more resilience towards both ends.

9.3.2 Implementation and Research Projects and Farm Water Management Plans Under DROP

An important report for drought policy in Twente has been "Sturen op basisafvoer" (2012, Steering for basic flow) for which the water authority cooperated with

Deltares (an important water management practice oriented research institute), the province of Overijssel, the farmers union (LTO-North), the drinking water company (Vitens) and Landscape Overijssel, the nature organization. In this report the problematic is analysed and a score of potential measures in the water system identified. The DROP pilot projects in the Twente region of Vechtstromen are related to the kind of measures proposed in the report. They consist of seven local *implementation projects* with an impact on the water system and two investigations. Also, two *research projects* were part of the activities in DROP. Finally, *water management plans* were developed with 15 farmers. The projects are carried out in the programme "Water Collective Twente". In fact this programme is a follow-up of two similar previous programmes: "Back to the source" and "Upgrading Water Management for Agriculture". Also in these programmes, like in the present one, collaboration and exchange with neighbouring water authorities took place.

9.3.2.1 Implementation Projects

Several implementation projects were realized in the northeast of Twente: drainage systems were removed, ditches were muted (sometimes with sand nourishment, see Fig. 9.2), streams were shoaled and water storage areas were constructed. Drainage systems are typically geared towards getting rid of the water as soon as possible. Deep ditches and creeks have a similar effect of nearby land: they extract groundwater from the shores with also some effect on ground water levels further

Fig. 9.2 Sand suppletion Snoeyinks brook

away from the water streams. Water storage basins on the contrary give surface water ample time to make it into the ground water, and additionally provide service water in dry periods. Due to these measures, it is expected that the groundwater level will rise, creating a water buffer for dry periods.

An example is the restructuring of the upper reaches of Snoeyinks brook. The project area comprises a number of small upper tributaries that flow into the Snoeyinks brook. The restoration of morphology and historical course of the river; the creation of new natural areas (hornbeam, oak woodland and poor-quality grassland); landscaping; recreational development, with access; and water management measures on farms and fields all contributed to making the area more drought resilient, while also improving the recreational and natural value of the area. The area is the property of Natuurmonumenten ("Nature Monuments"—Dutch Nature Preservation Society), farmers, and other businesses nearby.

Another example is the restructuring of the upper reaches of Springendal brook. This area is property of the State Forest Service (Staatsbosbeheer), nowadays actually more an NGO than a state agency. Here, the project consisted of making the brook bed shallower over approximately 300 m and altering its profile in combination with the restructuring of the landscape.

The selection of these projects has been done quite pragmatic. On the one hand they were identified with the help of a "desiccation map". But this just led them to contact relevant stakeholders there. On the basis of these stakeholder contacts, they selected the plots where they could expect most support and collaboration to develop pilots, being fully aware that being able to develop collaboration is the "name of the game". This way, nice examples are created that later act as marketing to attract other land owners to volunteer. Also it is important to look from the other side: which organizations have plans that are among others also helpful for drought resilience. Joining, supporting and modifying such plans could be a very good way to achieve goals, sometimes even better that starting with own plans.

9.3.2.2 Research Projects

Apart from implementation projects also research projects were part of the DROP pilot in Vechtstromen. One of these projects involved the testing of a level-dependent drainage system near a nature conservation area. Level-dependent drainage implies that the land owner or tenant can to some extent influence the water table by adjusting the drainage system. The advantage is that the table can be temporarily lowered when for instance the farmer wants to work on the land with machinery. This improves the farmer's preparedness to accept higher water tables than they would without a system they can influence themselves.

Many water managers see level-dependent drainage as the primary means of preventing water depletion and of optimizing agricultural use of areas of land. However, this idea lacks scientific underpinning. There is only limited knowledge about the effects that level-dependent drainage has on nature. Therefore, we conducted a study in a nature area that is surrounded by water-depleted area of

Fig. 9.3 Research equipment to study surface runoff

intensive agriculture, through which we aimed to better serve both nature and agriculture during long periods of drought. Vechtstromen water authority has constructed a system of level-dependent drainage in combination with raising the drainage basis of a small water-depleted nature conservation area to be able to test the effect on nature.

Another research project has been studying surface runoff. Surface runoff in Northeast Twente is commonly observed in hilly areas, where soil layers with low permeability reach the surface (Fig. 9.3). Extreme precipitation events, together with impermeable layers at shallow depths, can result in pool formation. If precipitation events follow shortly after land fertilization, pools with high phosphate concentrations are formed. As a result, the surface water is enriched with phosphates, leading to eutrophication. Research has been carried out to estimate the potential to improve drought resilience of a number of possible measures to reduce runoff. Examples of such measures are contour ploughing and the construction of earth banks along the low parts of fields to enhance water infiltration.

9.3.2.3   Water Management Plans

Water management plans have been made together with in total 15 farmers, tailored to their specific situation. The plans include tips and tricks on how to influence the water balance by storing water, resulting in a mutual gain for the farmer and the adjacent nature areas. The aim of these plans is to work on drought adaptation on a small scale, fitting in an overall vision for the area. The intensive communication established with the farmers created awareness, and motivated other stakeholders to

Fig. 9.4 Removed drain tubes

work on drought adaptation as well. One of the measures taken in consultation with farmers is to remove drainage systems that are in fact making the land vulnerable to droughts while lowering the water table too much (Fig. 9.4). The size of the "snowball effect" has surprised the practitioners. Several farmers have already volunteered to be included in a next round. Especially the smaller traditional farmers need to see results like better crop growth at the neighbour's plot, to get interested.

9.3.3 Drought Resilience Projects as Social Interaction Processes

All activities require a lot of stakeholder consultation and designing agreements with them. This is not only true for the water management plans that are designed together with farmers on a voluntary basis, but also for the implementation projects that are done in consultation with among others the municipalities, representatives of the provincial government, nature NGO's and other stakeholders. Even the research projects require extensive consultation and agreement.

All in all this does imply that the challenges for the practice projects are often not only technological but for a very important part also about the management of cooperation between various public and private stakeholders. The inclusion and participation of private actors such as farmers in the execution of measures, for instance in the level-dependent drainage approach or in the tailor-made water management plans, aids in aligning all actors on the awareness, importance and implementation of drought adaptation. It is essentially this management of cooperation in order to enable the realization of practice projects that requires a good governance context. Without a good governance context the degree of trust, openness and mutual liking is likely too low to allow for real cooperation. To what degree such a stimulating governance context is present in the Twente case, and what are the strong and weak aspects therein, will be the focus of the next section. In that section we will apply the governance assessment approach explained in the introduction.

9.4 Governance Assessment: Actor Coherence Saves the Day

The quality of the governance context and the way practitioners deal with that context is an important consideration to be taken into account when implementing measures for drought resilience. In the DROP project the governance context has been studied not only by reading relevant documents, but also by two visits of the DROP governance team. Interviews and meetings with the representatives of all relevant stakeholders during those visits, as well as studying the secondary data, provided a clear picture of the governance context. This enabled to assess it along the four governance criteria of *extent, coherence, flexibility* and *intensity*. The essence of those criteria will be repeated each time before describing the observations done in Twente region. The focus of the analysis will again depart from the perspective of the realization of the projects in the Twente region. That does not preclude that often policies and actors that operate at a higher scale will be mentioned and included in the analysis. But this is than always because of their relevance for the Twente pilot situation.

9.4.1 Extent: Are All Elements in the Five Dimensions that Are Relevant for the Sector or Project that Is Focused on Taken into Account?

The extent aspect of the governance context in Twente can mostly be regarded rather positive, but with some restrictions. It is positive in terms of the large extent of the levels and actors involved and high degree of openness of involved stakeholders, as well as awareness and increase of the visibility of the drought issues.

Regarding levels and actors involved we looked at the full range from the European level to the local pilot level. At the European level the Water Framework Directive and Natura 2000 policies make the relevance of the EU obvious.

The reform of the EU CAP[1] could be also relevant in the future. The national level seems to gradually withdraw from the process, even though at the same time the discussion on "fresh water supply" is gaining more attention. This is mainly due to the fact that the relation of water management with nature development has been severely damaged in 2011, when nature policy changes took place that almost completely cut the budget for new nature projects. This setback has never been completely restored. Other potential actors that are relatively absent are the drinking water company and the general public. Nature organizations are welcomed but are sometimes limited in their participation in decision-making due to restricted means.

In terms of problem perceptions about droughts, a gradually increasing number of included perspectives are observed. On a national level the problem of water scarcity is clearly addressed in the second Delta Programme, be it that the topic therein is "fresh water supply" which not necessarily leads to system adaptation to droughts. The visibility of this problem is not only present in the water and nature sectors but also gradually increases in the agricultural sector. Now, even in Spring, sometimes a few creeks and brooks run dry and more impacts on vegetation occur. This is also seen from the involvement of those sectors in the pilot area projects in the northeast and east of Twente region. The interconnection of drought and flood protection measures is increasingly recognized as having climate change as a common cause, which makes it somewhat easier for both problems to be addressed together. However, still the water authority itself in the organization where draught awareness is strongest. Already since around 2008 they have a permanent "draught team" in the organization.

A wide variety of instruments and measures is used, but as far as preventive measures are concerned they are restricted to a voluntary approach strategy. A specification of all desired water levels and tables serves as a basis for further extractions, especially in relation to Natura 2000, and as a guideline for day to day management. Around Natura 2000 areas buffer zones can be specified to protect the nature from lower water tables and chemicals used in the agriculture. A new instrument included in the ZON agreement mentioned in Sect. 9.2 is the specification on how much water farmers can expect during wet or dry periods. This should enable farmers and industries to consciously take or avoid risks for instance with high value crops. Moreover, the obligation to create a storage capacity for a 20 mm rainfall in case of a new building or new development decreases the amount of rainfall to get in the sewage system, which not only prevents flooding in the cities but also prevents the ground water level to drop. This enables more infiltration and watering the street trees by stored water in dry periods. Also, other instruments such as a ban on irrigation from surface water in certain dry periods and a ban on extraction of ground water in certain areas imply a growing awareness to increase

---

[1]The CAP reform of 2009 introduced 2 new standards of GAEC (Good Agricultural and Environmental Condition) related to water: (a) establishment of buffer strips along water courses, (b) compliance with authorization procedures for use of water for irrigation. Retrieved from: http://ec.europa.eu/agriculture/policy-perspectives/impact-assessment/cap-towards-2020/report/annex2a_en.pdf.

drought resilience. The water authority has tried to harmonize its regulations regarding irrigation policy and permits for water extraction with the neighbouring water authorities.

Though a multiplicity of relevant responsibilities and resources resides with many stakeholders, it is mostly the water authority that clearly frames them in the view of drought resilience. Formally, the province has the responsibility to set the goals for water management. The provincial domain also includes responsibilities in relevant sectors like nature protection, nature development, landscape and tourism, regional development and agriculture. The water authority implements these, but with an own domain of taxation and thus with considerable liberties. The water board also has a lot of specialized knowledge that the other actors need for their decision-making. On the level of pilots and in the projects with farmers there is some restriction of instruments to voluntary approaches as well as lacking tax incentives for farmers as a mechanism to promote drought resilience. There is a big variety of relevant responsibilities and related resources with involved stakeholders. However, most responsibilities in relation to drought resilience are with the regional water authorities.

9.4.2   Coherence: Are the Elements in the Dimensions of Governance Reinforcing Rather than Contradicting Each Other?

The coherence aspect has a similar positive assessment, though it is more complex. At the levels and scales dimension the relationship between the levels is observed as a soft hierarchy by multilevel agreements. The unique multilevel coordination in water policy in the Netherlands is guided by so-called National Administrative Agreements on Water in which all levels from national to local and from government to (drinking water) companies mutually agree on their share of the tasks, responsibilities and funding. While the recent OECD Dutch water governance assessment speaks of the lack of "independent oversight" at the multilevel dimension, there are mechanisms such as multi-stakeholder Delta Programming and the National Administrative Agreements on Water Management with monitoring strategies. These should not be underestimated as mechanisms to provide multilevel coherence without coordination from a central power.

In a similar way in the pilot area all four local authorities, the water board and the provincial government together made an Area Vision, providing a joint perspective on desirable developments of the northeast area of Twente. Though drought resilience is included, the main focus is on tourism and recreation. Because of these collaborative efforts, nature organizations are more open than in the past for consultations and compromise. The drinking water company has regular contacts with the farmers union LTO (when trying to find new locations for wells), who reports that the collaboration with other partners, including the water board and nature organizations, has significantly improved during the last two years. Being part of the process creates also gradually more enthusiasm among the farmers. Agricultural interests can this way become coherent with the interests of drought resilience projects that often imply forms of re-naturalization. This close collaboration of multi-stakeholders at levels of administration and project managers is a great advantage that provides coherence and enables successful implementation of measures. It can also be seen as a necessary and relatively successful adaptation to deal with an inherent rather than incoherent and fragmented governance context, labelled as the fragmentation—coherence paradox. It is called a "paradox" because, while normally fragmentation would lead to stalemates and ultimately disinterest in the topic, in a context of sufficient positive experiences with mutual cooperation it has led to a recognition that the various parties need each other and to the absence of fear that one of them will become too dominant. This "being part of the process" creates enthusiasm with farmers. Trust between farmers and the government is extremely important. Otherwise also mutual social control among farmers will turn against collaboration or even selling agricultural land to the government.

In terms of problem perceptions, drought resilience is not yet a fully shared priority. Instruments and strategies are not balanced and relatively fragmented in their consequences for implementation. As for responsibilities and resources, they are fragmented in a very complex way that only mutual consultations are the ways to proceed, creating the already above mentioned fragmentation—coherence paradox.

9.4.3   Flexibility: Are Multiple Roads to the Goals, Depending on Opportunities and Threats as they Arise, Permitted and Supported?

The flexibility aspect gets a moderately positive assessment with fair degree of adaptive capacity. It is for instance seen in the way how the province and the municipalities took over on the emptied role of the national level, showing a healthy degree of flexibility. The high interconnectedness character of the group has advantages to get political and financial support for projects and thus contributes to flexibility.

However, there are sometimes inflexibilities that derive from the geo-physical and landscape conditions of the small-scale Twente area, or relate to the strong Dutch local land use planning system. The Dutch planning system has lengthy decision-making procedures, like EIA, not only in the Twente region but in the whole country. For example, the creation of a new well for drinking water production has to undergo a lengthy procedure even when aiming to replace a well that has more impact on the drought resilience of an area. Also the stakeholder representation by institutionalized organizations can sometimes lead to inflexibilities. For example, not in all cases it is easy to get quick recognition as a new actor in the process, as was reported by a farmer within a group operating independently from the farmers union LTO.

Moreover, the restriction to voluntary approaches for preventive measures also makes the governance context somewhat inflexible. The general strategy of the projects is to convince or inspire new groups of farmers to join in voluntary drought resilience projects and by that having an opportunity to avoid obstacles. Drought resilience policies should have an impact on spatial planning as Dutch local spatial plans are relatively inflexible. It is the collaborative relationships and coherence of the actor groups combined with high level of trust that enable to pool goals, instruments and resources in such a way that a reasonable positive degree of flexibility can be assessed. However, such flexibility could be restricted with the development of more specific accountability regulations in European and national subsidy schemes. This has already happened in river restoration cases in the Twente region. A restriction to the development of long-term pooling of resources is furthermore the increased emphasis on the "innovative character" of the projects proposed. This seems at the surface to promote flexibility, but in fact might turn out to prevent successful pilot measures to be developed into the implementation of large area scale projects that really make a difference.

9.4.4 *Intensity: How Strongly Do the Elements in the Dimensions of Governance Urge Changes in the Status Quo or in Current Developments?*

Intensity quality appears to be the weakest point of the governance context for drought resilience policies in Twente. There seems to be no political support for forceful measures, but only for voluntary ones in the preventative sphere. The budget cuts in the sectors of nature development and landscape protection contribute to that. The withdrawal of the national level in the nature policy could have had serious effects for the funding of the projects but luckily the province took over on a large part of it. The preference of the province is to concentrate on the position of Natura 2000 areas. This has implications for to the funds and permissions for the projects in the pilot area as some that are closer to Natura 2000 areas are easier financed than other, even though they might be equally well serving agricultural drought resilience. Also, the WFD directive provides pressure to make water systems more resilient. These two external pressures make the overall assessment of the intensity as medium.

After the national government stopped the National Landscape programme, and thus also the northeast of Twente had no longer that official status, the provincial government kept treating the northeast Twente as one. Since most practical activities in the framework of the landscapes are related to recreation and tourism development, municipalities play often a leading role in this respect. The water authority as well assumes co-responsibility for nature but only as far as water goals are addressed and a good water system is needed to support the nature.

Actor and network' intensity reflects the mixed priorities of these stakeholders. The realization of drought resilience is seen as a clear priority for the water authority. This makes them a lead actor, even when they involve other stakeholders as well in their initiatives. With regard to farmers, their position has considerably developed in the last few years. The farmers union LTO now wishes to improve awareness on drought issues to better educate farmers. Moreover, while the EU milk quota are lifted in 2015, it becomes worthwhile to invest in the quality and resilience of the land. Individual farmers are willing to cooperate when good projects are offered, even when the farmers union was still hesitant.

Almost all actors, except the nature organizations, share a preference for voluntary actions. For the farmers freedom of crop choice and increasing productivity are essential for their support for drought resilience measures. There is a Deltaplan Agrarian Watermanagement (LTO 2013) in which the LTO seeks to collaborate with water boards, provinces and ministries in achieving water goals, including a 2 % annual production growth among the objectives. Municipalities do also collaborate but for them often drought is not regarded as a prime problem. Municipalities experience pressure like budget cuts due to decentralization.

While broad awareness of the drought problem perception is just developing, external legal pressures stay dominant. The drought problems are taken more seriously after some recent examples of spring droughts that have more serious impacts on yields, than summer or autumn droughts. External legal pressures from Natura 2000 requirements for the designated areas provide a strong stimulus for drought prevention measures regardless the economic interests and other values, challenging the farmers to push the boundaries beyond the "business as usual".

For farmers drought problems are less visible than flood risks. There is a tendency that crop damage results in relatively higher prices so that only farmers that have been hit in an exceptional degree are disadvantaged. Nevertheless, the farmers union is now convinced that after working with individual farmers a new approach in which measures are taken in larger areas at a time is needed.

Whether voluntary projects will ultimately provide sufficient incentive to enable continuous improvements in drought resilience in the pilot area and elsewhere is an open question. Due to a lack of strong political support and legal pressures from the Dutch national level, where drought issues are still just at the beginning of broad recognition, this question is not really debated. In addition there seem to be no viable alternative options given the division of responsibilities and resources.

It is hard to say whether the resources will be sufficient for the drought resilience goals. The province has decided to invest some 330 million Euro in increasing the buffer zones around Natura 2000 areas. They also took care for replacing national funding for part of the ecological network EHS and for the Area Vision for northeast Twente. The farmers union is given a privilege by the province to be the main implementer of the programme to enlarge the buffer zones around the Natura 2000 areas to increase the drought resilience. The water authority is also prepared to invest its resources. The nature organizations feel themselves very limited financially and understaffed. They sometimes cannot participate in consultations where in principle they would be welcome to have an input. This situation has worsened over the years. To end with, it is an issue whether drought resilience policy could or should go beyond voluntary approaches.

9.4.5 Overview and Visualization of the Results of the Analysis

Figure 9.5 attempts to visualize the results of the governance context. By strongly summarizing the original assessments and their explanation above, this obviously implies the loss of a lot of nuance. Nevertheless, the figure has also an advantage, namely that is provides an overview. This shows for instance that concerning *extent* especially the "levels and scales" box is relatively weak, this being largely the result of the withdrawal of the national level from relevant policies as nature and land-scape. With the criterion of *coherence* the column illustrates the "fragmentation-coherence paradox": while fragmentation is present in all boxes but one, especially with the "responsibilities and resources for implementation", it is the excellent (*dark* green) situation in actor coherence that saves the situation. *Flexibility* is quite good, though the low degree of alternative sources of income next to market oriented farming in the "problem perceptions and goal ambitions" box and the sometimes rigid land use planning in the "strategies and instrument" box are a bit less supportive. Finally, the *intensity* column looks most gloomy. Especially the slow integration of the drought resilience awareness and the resulting reliance on voluntary preventive measures only create the risk that the ultimate goals will be difficult to achieve.

Dimensions	Criteria			
	Extent	Coherence	Flexibility	Intensity
Levels and scales	+/0	0/+	-	0/-
Actors and networks	-	++	-	0/+
Problem perspectives and goal ambitions	-	0/+	0	0/-
Strategies and instruments	-	0	0/+	0/-
Responsibilities and resources	-	-	-	0
	Colours **red: restrictive;** *orange: neutral,* *green: supportive*			

Fig. 9.5 Visualization of governance context assessment conclusions (*the darker green* implies intensely supportive)

The observations described above conclude that the governance context for drought resilience policies and measures for the Twente part of the water authority of Vechtstromen can be regarded as moderately positive (supportive or at least neutral), though obviously such general conclusion is always relative to other situations and dependent on the choice of issues emphasized most.

9.4.5.1 Contextual Interaction Theory

Following Contextual Interaction Theory such a generally supportive governance context is expected to have positive consequences for the motivation, cognitions and resources of the actors involved in the process of implementing drought resilience measures. What we did observe is a varying and on average moderate, but also growing degree of draught awareness among the actors. This also positively influenced their motivation. Of course there is here a strong pre-selection effect: while all measures are taken voluntarily it is no surprise to see predominantly positive motivations. On the other hand: also with previously sceptic actors like the agricultural organization positive developments are observable. The resources of the actors made available for drought resilience measures and to push and pull to get them accepted are generally sufficient, with the exception of the withdrawal of the national government from the support for nature development and the national landscape park that has made things more difficult. On the positive side, the provincial government has compensated this loss to a large degree, indicating its preparedness to contribute. All in all this has led to an implementation process with a remarkable degree of collaboration.

When we look more specifically into the three actor characteristics of Contextual Interaction Theory we see the following. The *motivation* of the stakeholders working to act on drought is triggered to large extent by their own goals and values. In a dense country such as the Netherlands, the last remnants of nature are seen as especially precious. In addition, desiccation is considered one of the major environmental policy themes in the Dutch government for several decades now. The local water authority views cultivating drought resilience for the area as its own responsibility. Also nature organizations and more and more farmers are aware of the implications of drought and water scarcity, including loss of flora and fauna in creeks and crop losses. For the province also the external pressure of European policies is very relevant. Low self-effectiveness assessment on top-down regulation of preparatory measures limits the scope of instruments to voluntary ones.

The *cognitions* of actors are mainly driven by observations on the changes of the regional water balance. The visibility of drought issues has increased in the region, with creeks running dry as well as dry vegetation. Damages from droughts are affecting agricultural yields in rural areas. Cities and their urban infrastructure are also affected by drought, which in leads to a change in awareness and perception.

In contrast to the strong legal requirements behind nature conservation in the Netherlands there are limited *resources* for the nature organizations to combat drought issues themselves. Their key competencies are rather limited. Other actors, such as the water authorities, have much more flexibility to take over the task of drought resilience improvement. Nevertheless, also in their case the capability to enforce resilience measures is restricted to a voluntary approach, partly because of limited legal possibilities, partly because of lack of political support for coercive action outside emergency periods.

9.5   A Tale of Preserving Voluntary Action and Upscaling Nonetheless: Conclusions and Recommendations for the Region Twente

The conclusions and specific recommendations below are partially based on comparing the Vechtstromen context with Governance Team members' knowledge of other water management systems, including a comparative analysis with the other regions studied in DROP. We will concentrate on the preventive measures for increasing drought resilience that are central stage in the most vulnerable areas of Vechtstromen like the northeast of Twente.

9.5.1   Overall Conclusion

Nationally the recognition of the problem is still at an early phase and a water supply orientation is still dominant. Against this background there is only legitimacy for soft voluntary approaches to prevention policies and measures. Partly this approach is also rooted in the general Dutch consensual political culture (the so-called "polder model"). This forms a setting in which building and using well-functioning partnerships with as many stakeholders as possible, both allies and potential opponents, is the best way to make the most of the situation and create the best likelihood of success. We observe that the project managers of the water authority understand this very well and are doing a good job at realizing it this way.

9.5.2   Awareness and Public Agenda

Drought and fresh water shortages being still a low profile issue has the disadvantage that financial and political support for preventive measures to increase drought resilience is limited and the relative priority of such measures is weak when they compete with other objectives. A background is that the most populated parts of the Netherlands have an artificial water system that allows to manage water levels to prevent droughts by bringing water from other areas. On top of that water scarcity is often related to the prevention of saltwater intrusion by using large quantities of freshwater to flush these artificial waterways. All of this makes the problematic of the "higher sandy soil areas" where dependency on rain water and ground water increases the risk of drought damage for both nature and agriculture. Thus, continuous efforts to get and keep this issue on the national agenda are warranted. Because large parts of the south of the country are facing exactly the same problematic, it is recommended to do this in close collaboration with not only the eastern, but also the southern water authorities.

9.5.3   Inter-collegial Exchange and Learning

As a consequence of the relatively low saliency of the drought problematic, but also related to the objective of the water authority of Vechtstromen to maximize the value of water measures for a broad range of societal goals, the drought resilience projects in the northeast of the Twente region are not regulative in character, but supportive, voluntary and consensus oriented and aiming for the integration of various sectoral goals. In our conclusions we already stated that this is given the governance context likely the most efficient way to proceed and also that the project managers are doing a good job at this. However, the period that these projects will take to increase the resilience of the whole vulnerable area might be so long that from time to time new or extra project managers will be involved. The other way around: the expertise and experience gained from the drought projects in consensual project management could also be beneficial for project managers in other water projects. For these reasons we recommend to organize venues for inter-collegial exchange and learning, for instance by regular sessions.

9.5.4   From Farm Level Approach to Full Area Level Approach

The extension of local drought resilience projects wherever good chances for realization have occurred and stakeholders could be convinced to cooperate had led to a wide array of very nice projects. The basic idea is that each generation of projects will convince other stakeholders to participate in a next round and thus create a bandwagon effect. This way many local project areas have been improved in terms of drought resilience. This approach has great virtues and should be continued. However, this will not necessarily lead to the full coverage of the water system in somewhat bigger areas, for instance watersheds of creeks and small rivers. Therefore, it might be necessary to add to the small-scale voluntary project approach a new sort of approach. In this approach somewhat bigger areas than separate farms could be the scale, for instance containing one or two dozen farms of which for instance two have been already involved in showcase projects in previous rounds, attempting to make the coverage by drought resilience measures for that area complete and create synergy between the measures.

9.5.5 Creating a Long-Term Outlook and a Vision for Each Area

When many stakeholders have to align in multi-purpose (but in any case also drought-related) measures, it is important that the outlook for the vulnerable region as a whole and visions of the desired status of the subareas that should be dealt with are as clear as possible. Vechtstromen has had a good experience with this way of working when long-term complex and dynamic implementation process should gradually realize a better status of the water system. An "iconic" example is the restoration of the Regge River (the main water course of the Twente region) (De Boer and Bressers 2011), but also the reconnection of large parts of the region to the Regge River by realizing a new river, de Doorbraak (the "Breakthrough") (Bressers et al. 2010) and the present works on the Vecht river are examples. The "compass" effect of such an integrated multi-sectoral vision on both the scale of northeast Twente and more specifically subareas with their own characteristics is a proven requirement for such long-term efforts and should also be further developed in this case.

References

Bressers H, Hanegraaff S, Lulofs K (2010) Building a new river and boundary spanning governance. In: Hans B, Kris L (eds) Governance and complexity in water management. Edward Elgar, Cheltenham, pp 88–113
Bressers N et al (2015) Practice measures example book DROP: a handbook for regional water authorities. Water authority Vechtstromen, Almelo
de Boer C, Bressers H (2011) Complex and dynamic implementation processes. Analyzing the renaturalization of the Dutch Regge river. University of Twente and Water Governance Centre, Enschede, The Hague
Deltacommissie (2008) Samen werken met water (Working together with water). The Hague
Deltaprogramma zoetwater (2013) Kansrijke strategieën voor zoet water: Water voor economie en leefbaarheid ook in de toekomst (Promising strategies for fresh water). The Hague
Kuijper MJM, Hendriks DMD, van Dongen RJJ, Hommes S, Waaijenberg J, Worm B (2012) Sturen op basisafvoer (steering for basic flow). Deltares, Delft, Utrecht
LTO (2013) Deltaplan agrarisch waterbeheer (DAW) (Deltaprogramme agrarian watermanagement). LTO, Zwolle
OECD (2014) Water governance in the Netherlands. Fit for the future? OECD, Paris

Chapter 10
Cross-cutting Perspective on Agriculture

Gül Özerol and Jenny Troeltzsch

10.1 Introduction

Agriculture is among the major water user sectors in the North-west Europe region. While the use of water for energy production, mainly for cooling of power plants, has a higher share in most of the countries in the region, agricultural water use maintains an average of 24 % share within the total water use in Europe (EEA: European Environment Agency 2009). At the same time, agriculture is a vulnerable sector in terms of the impacts of drought on both global and European agricultural production (Geng et al. 2015). Although making a universal definition of drought is not straightforward due to its diverse drivers and impacts, the following definition can be made for agricultural drought: "the result of a shortage of precipitation over a particular timescale that leads to a soil moisture deficit that limits water availability for crops to such an extent that yields are reduced" (Sepulcre-Canto et al. 2012: 3519). A key relationship that is addressed in this definition is the sensitivity of crop yields to limitations in water availability.

This chapter elaborates on the governance of drought adaptation in the North-west Europe region from an agricultural perspective. For this purpose, the elements of the governance systems that are relevant for agricultural production and water use processes are examined and their influence on drought management and adaptation processes are investigated in the subsequent sections of the chapter. In each section, illustrative examples are provided from the six cases of the DROP project that are presented in Chaps. 4–9 of this book. Section 10.2 starts with the general problem

G. Özerol (✉)
CSTM—Department of Governance and Technology for Sustainability,
University of Twente, 217, 7500 AE Enschede, The Netherlands
e-mail: g.ozerol@utwente.nl

J. Troeltzsch
Ecologic Institute, 10717 Pfalzburger Strasse 43/44, Berlin, Germany
e-mail: jenny.troeltzsch@ecologic.eu

© The Author(s) 2016
H. Bressers et al. (eds.), *Governance for Drought Resilience*,
DOI 10.1007/978-3-319-29671-5_10

perspective by describing the drought and water scarcity problems related to agriculture. Section 10.3 examines the intersectoral linkages by examining the relationship of drought with the risks of the competing sectors of agriculture. Then Sect. 10.4 focuses on the multiplicity of governance levels and outlines the interactions among local, regional, national and European Union (EU) policies and actors. In Sect. 10.5, awareness on the agricultural impacts of drought within the public and policy spheres is assessed. Finally, Sect. 10.6 synthesizes the discussions presented in the previous sections in order to provide an outlook regarding the adaptation to the existing and future impacts of drought from an agricultural perspective.

10.2 Drought and Water Scarcity Problems Related to Agriculture

In many areas of the world, drought negatively affects both rainfed and irrigated agriculture due to decreased water availability and quality. As a result, two common impacts of drought on agriculture are often observed, namely decreased crop yields and harvest qualities. In the North-west Europe region, these impacts result in several implications for the agricultural practices and the farming community, as illustrated in the six case study regions. In the paragraphs below, the general problem perspective in each case study region is described first by providing the role of agricultural production and then by explaining the recent drought occurrences and their implications for the agricultural sector.

More than 46 % of **Flanders**' surface area is used for agriculture and counts for 1.5 % of the gross domestic product of Flanders. The total area of land earmarked for farming has remained roughly the same over the last years. The farmland is mainly situated in the provinces of West and East Flanders, in Hesbaye and Northern Campine. 56 % of the agricultural area is covered with fodder crops (meadows, pasture and feed maize), which can be explained by the importance of stockbreeding (mainly pigs, cattle, poultry). Arable farming uses 35 % of the agricultural area. The main crops are cereals, potatoes and sugar beet. On 8 % of the agricultural area horticulture is practised, mainly for vegetables and fruits. Horticulture is a very relevant economic area for Flanders, e.g. Flanders is the world leader for export of frozen vegetables (Platteau and Van Bogaert 2014). Drought issues are not widely discussed in the Flanders' agricultural sector. However, for horticulture a high water quality is necessary, which partially cannot be reached at some times during the year. Furthermore, the sandy loam area in the centre of Flanders accounts for the high-value horticulture production, but is very dependent on rainfall and groundwater. In the past, Flanders has experienced droughts in the years 1976, 1996, 2003, 2006 and 2011. In recent years, droughts have had several consequences in Flanders and on several occasions water extraction from the Albertkanaal has been restricted. The drought period in the summer of 2003 did not reduce the agricultural yields. This was attributed to the fact that the drought was not severe enough in the growth season to have had a significant impact (UN 2004). The 1996, 2006 and 2011 droughts were recognized as agricultural disasters, and affected farmers were financially compensated (Chap. 7, UN Department of Economic and Social Affairs Division of Sustainable Development 2004). The expected decrease in summer precipitation—coupled with a possible increase in summer water demand due to higher temperatures, in particular if irrigation becomes a widespread agricultural practice—can lead to a further lack of water availability and problems for the agricultural sector in Flanders (UN Department of Economic and Social Affairs Division of Sustainable Development 2004).

Among the different administrative districts in the **Eifel-Rur** area, the northern downstream area is more characterized by agriculture. The administrative district of Heinsberg, which covers the downstream area of Eifel-Rur is with 65 % the district with the highest percentage of agriculture area in the County Cologne (Bezirksregierung Köln 2013). The southern upstream part of the Rur basin is an area with a low population density, so that for all administrative districts in Eifel-Rur agriculture shows about 1–1.5 % of their gross domestic product. The neighbouring administrative districts have about 0.5 % of their gross domestic product in agriculture which shows that agriculture in Eifel-Rur is relatively important (Landwirtschaftskammer Nordrhein-Westfalen 2012). For the County Cologne, the agriculture area is mainly used for cereals, root crops (e.g. sugar beets), maize and fodder. Stockbreeding covers mainly cattle and poultry. Past drought episodes with consequences for agriculture in the area are very limited. Also the possibility for irrigation is low—with ca. 5 % of the agricultural land. As an example, in 2009, only half of the land which has irrigation infrastructure was actually irrigated (Landwirtschaftskammer Nordrhein-Westfalen 2012). However, in the lower downstream area negative water balance during dry summers is seen as a problem for agriculture, although with a low intensity.

In the **Salland** region, main water use for agricultural purposes is the irrigation of crops and grass. Both groundwater and surface water are used for irrigation. In the 1980s, drainage and groundwater extraction for irrigation and drinking water caused desiccation, which damaged the agricultural areas and the nature areas that were sensitive to the changes in groundwater level. As a result, irrigation with groundwater was banned in the 1990s and 2000s when the water sufficiency was threatened for drinking and industrial uses and for nature areas. However, some agricultural areas have already become dependent on irrigation. This dependency implies that the agricultural areas can dry out without irrigation, which can lead to significant decreases in agricultural production. The vulnerability of these irrigated agriculture practices is expected to worsen also as a result of the increasing pressure for protecting the nature areas that are sensitive to groundwater levels.

The **Vilaine** catchment is a rich agricultural area, where tourism, industry and navigation are also among the major economic sectors that demand water. Main agricultural crops grown include cereals for animal feeding (Bouraoui et al. 2009), and market-oriented gardening, such as cauliflowers, in the eastern part of the catchment. A diversity of agricultural profile can be pointed out all around the catchment. In 1976, 1989 and 2003, severe droughts were experienced in France, which also influenced the Vilaine region. All of these droughts had significant damages on many sectors, including the agricultural production. As a result, several water- and agriculture-related measures were taken, such as financial protection for farmers and irrigation bans.

Agriculture and food and drink production are major industries in the **Somerset** County. Somerset is a major producer of cider, based on their apple orchards. Furthermore, farming of sheep and cattle and the production of cheese are important in the region. In the whole south-west area of England two-thirds of the land is devoted to agriculture, which employs 3.7 % of the workforce in the region. In 2010–2012, Somerset experienced a drought event with consequences for the whole region, also farmers. Furthermore, expected wetter winters and drier summers are likely to have a profound effect on land management and farming. Building soil organic matter is crucial to drought-proofing soils in Somerset. It is expected that the cultivation of new crops such as grapes, maize, sunflowers and soya will increase in the region, with the consequences of an increased need for irrigation, owing to reduced summer rainfall and higher temperatures. Over the coming century, the region's water resources will come under greater strain as summer droughts potentially grow longer and the demand for irrigation increases.

Main crops grown in the **Twente** region of the water authority Vechtstromen are grass and corn for animal feed and high-value crops, such as flower bulbs and potatoes, all of which are water sensitive. Thus the agricultural sector in Twente is also vulnerable for water scarcity and droughts. Irrigation bans were announced in the 1990s, partly as a result of drainage measures that were taken for the wet periods. Grassland farming is expected to be intensified, and thus need more water for irrigation, which is likely to cause more irrigation bans during dry periods. The recent irrigation policy, which was adopted in 2013 and applies to the areas of all the water authorities in the eastern Netherlands, including the Salland region. The policy aims to balance the use of groundwater and surface water by farmers and the water needs of vulnerable nature areas. However, the policy will only apply to groundwater extractions from new wells, whereas the existing wells are excluded, thus reducing its potential impact .

The agricultural sector in the case study regions can be characterized as being sensitive to water availability, mainly in irrigated areas, and thus negatively affected from the past occurrences of drought. Eifel-Rur region is the only exception to this characterization, where drought has very low impact on agriculture. Furthermore, in all the regions, during drought periods multiple water user sectors are prioritized and the agricultural sector often receives a lower priority than that of domestic, environmental and industrial uses. This lower priority implies restricted or reduced water availabilities for agriculture, and even irrigation bans during severe drought occurrences. The sensitivity to water availability is expected to increase due to varying drivers such as further intensification of farming, increased cultivation of water-demanding crops, decreases in summer precipitations and the need for longer irrigation periods.

10.3 Drought Issues and Competing Sectors' Risks for Agriculture

Balancing the water needs and demands of the agricultural sector with those of the other water user sectors is a challenging task in many countries. Achieving this balance becomes a greater challenge due to the risks associated with drought. As illustrated by the findings from the six case studies, the relative importance of the agriculture sector in the regional economy is a significant factor that influences the resources that are made available to the actors of agricultural production and water use. In all the regions these actors mainly include the farmers, their organizations, and the water authorities.

In **Flanders**, current discussions on drought and water scarcity issues generally include many different perspectives, e.g. from farmers, nature conservation organizations, drinking water companies, etc. However, especially groundwater issues is a well-developed topic in Flanders and the focus in these discussions is mainly on agriculture and economic developments, where as other perspectives such as nature conservation, are less integrated. Various measures were developed for groundwater, but problems still exist with instruments such as source protection (quality), which requires land use changes or change of agricultural practices. Here local authorities are not enforcing and implementing such measures, because they would affect the economic development in the agriculture sector. Advisory services for farmers seem to be quite developed and they work towards water saving. Via the EU Common Agricultural Policy (CAP), investments in more water efficient technologies, such as water reuse, are supported. Furthermore, farmers are taking initiative to build water retention basins on farm level, but the permission process is mostly quite lengthy because farmers need to prove in the application process that the basin will not in reality end up capturing groundwater.

In **Eifel-Rur**, the water board is not responsible for delivering water for agricultural use/irrigation. Therefore, industries and drinking water companies are members of the water board, but not farmers and farmers' associations. Because of this structure, the agricultural water users can not influence the discussions as much as other users. However, this structural problem is offset by the fact that there is a growing culture of exchange and collaboration with smaller stakeholders, such as farmers, which all sides see as a productive relationship that is developing positively over time. Farmers agreed voluntarily to contracts, e.g. addressing nitrogen use, and are therefore cooperating with national conservation organizations, especially the national park authority.

Agriculture is a key economic sector in the **Salland** region, and therefore balancing the water needs of the agriculture with other water uses, particularly the environment, has been and is a crucial objective for many stakeholders, including the water authority, farmers, province and nature conservation organizations. Historically, the risk of flooding has been felt more commonly by the farmers, thus leading to prioritizing the discharge of excess water, rather than water scarcity and drought. However, some measures are taken, such as the irrigation bans during times of drought to protect the drinking, industrial and environmental uses. Currently, there are no comprehensive measures to address the water scarcity and drought from cross-sectorial perspectives, such as the monitoring or metering of irrigation water use at the field level or the enforcement of the new irrigation policy also for the existing groundwater wells. Nevertheless, there is a growing awareness on the impacts of drought and other climate extremes.

In the **Vilaine** region, the demand from the agricultural sector does not constitute a significant pressure on other uses, except the eastern part of the upstream Vilaine, where irrigation and nature protection are the two competing water uses. Similar to the Salland and Vechtstromen cases, water withdrawals by agro-industries and farmers are not monitored, yet it is planned. Furthermore, the CAP impacts the agricultural water use as well as the consumption of different products, while the measures related to the implementation of the CAP do not address potential interactions with the water policy. However, an agro-environmental measure is funded by the SAGE and implemented by the IAV, through offering contractual measures over five years to farmers who manage wet meadows in the marshes and can receive financial subsidies to maintain the marshes.

In **Somerset**, there are cross-boundary issues that span drinking water supply, environmental flow, and agricultural water use, but planning activities are not coherent. The Environment Agency has a drought plan that covers both water supply and agriculture and irrigation that covers a region rather than a water company. But the water companies have drought plans which cover drinking water supply (in balance with other environmental factors like flow). A more integrated approach could improve the activities. Furthermore, in Somerset it is clear to all actors (it was at least before the flood) that water scarcity and drought is a problem and will probably increase. A difference was in the recognition of the extent of the water scarcity and drought issues and the measures which should be taken up. Problem perceptions were largely defined by sectorial interests, particularly agriculture and nature on the Levels and Moors. For the Levels and Moors there are no real mechanisms to persuade landowners to keep their stock out of the grass and to keep the water levels up for delivery of other ecosystems services other than agriculture and no reason not to intensify agricultural production in those fields. Higher tier agriculture Sites of Special Scientific Interest (SSSIs) focused on exclusivity but not interconnectivity, for example they are not linked-up special sites. Although there are designated SSSI areas there is seen to be not enough guidance for farmers outside these specially designated areas, and little opportunity to "enforce". The range of measures are largely positive, however, in the context of decreased regulation and public spending the extent to which these are monitored and enforced going forward is uncertain. Furthermore, the subsidy regime was not developed with the aim that farmers should directly manage water and a system of monitoring or enforcement related to this was not established. For example, issues such as soil compaction were seen as voluntary actions as opposed to embedded in subsidized actions. A fuller range of agricultural measures could still be implemented, e.g. agricultural mitigation measures and the adoption of clearer monitoring and enforcement.

In the **Twente** part of Vechtstromen, irrigation bans are applied during dry periods, yet little pressure is felt in the agricultural sector to incorporate comprehensive water scarcity and drought measures, such as metering groundwater and surface water extractions. On the contrary, the intensive agriculture practices, which imply, among others, increased use of machinery and irrigation water, have a negative impact on surface runoff and groundwater levels, respectively. There are several agri-environmental measures, which are mostly voluntary and fragmented, and thus can address water scarcity and drought limitedly. The political and public awareness is seen as too low to create incentives that combine drought resilience and agricultural objectives. Two types of intersectoral competition affect the drought resilience of the agricultural sector both in Salland and Twente. First one is on the difference regarding surface water use priorities in the east and west of the country. In the west part of the Netherlands, surface water needs to be flushed to prevent salt intrusion, which implies a lower availability for agricultural use in the east part. Second one is the displacement chain, which outlines the priorities of competing water uses during dry periods, when meeting the water demand for all uses becomes impracticable. As explained in Chap. 8, agricultural production has a low priority in this chain. Although both of these competitions exist for a long time, they can become threatening for the agricultural sector, under worsening water scarcity and drought conditions.

Looking at the intersectoral linkages related to drought and its agricultural impacts, a diversity of situations can be discerned from the six case study regions. On the one hand, in the Flanders and Somerset cases, several measures are in place to address the cross-sectorial issues and to encourage measures, such as water saving and reuse, with varying enforcement levels among the regions. On the other hand, in the Vilaine, Salland and Twente cases, the competition among the sectors are addressed, while the current high water availability levels do not create enough incentives to monitor water withdrawals in the agricultural sector. Finally, in the Eifel-Rur region, the relatively low importance of the agricultural water use leads to a lack of recognition regarding the intersectoral linkages, with a growing interest from the side of farmers.

10.4 Multilevel Interactions Regarding Agricultural Measures

Multiple governance levels can be relevant regarding the agricultural practices and the political and practical measures related to drought adaptation and management. Within the North-west European context, there is an inherent role of the EU-level policies on water and agriculture, whereas the local and national actors and policies also interact regarding the formulation and implementation of agricultural practices and measures.

For **Flanders**, all matters related to agriculture come within the scope of the regional authorities, apart from food safety, which is still a federal Belgium policy area. In the form it takes in Flanders agricultural policy which is essentially European based. At this level Flanders is able to shape its policy in the light of what is decided by the EU authorities. Drought is seen as a problem by some farmers but the uptake of the problem by different political level starts slowly. The regional Flemish environment agency (VMM) starts to implement some initiatives and measures, but provinces and municipalities have still a limited awareness of the potential problem. Beside this, provinces and municipalities include the possible consequences on agriculture and other economic activities in their decision-making. In general, both levels work closely together.

The two main actors that play an important role in the water management regime at the **Eifel-Rur** area are the water board (WVER) and the district government. The water users with a water right of a certain size are automatically members of the water board in the area. However, smaller actors such as farmers do not have the same voice because they are not water board members. During the implementation of the Water Framework Directive (WFD), a large stakeholder process was organized by the region's water board. Therefore, the actors that are not involved as water board's members are also integrated in the discussions and seem to be quite satisfied with the participatory process. Since, agriculture is a relevant economic activity in the downstream areas of the Rur basin, the farmers as a stakeholder group are in a position to impose their own agenda to a great extent. There seems to be a reluctance to collaborate with water management objectives (e.g. when measures do not coincide with agriculture aims). For instance, municipalities with strong farming presence would resist repurposing some areas of land for WFD Programmes of Measures, although the legal basis is clearly against them. Also for drought-related measures this kind of deadlocks can appear.

In the Netherlands, thus for the **Salland** and the **Twente** regions, a broad range of governance levels are relevant, whereas the regional level is the most prominent one due to the role of water authorities at this level. Through designing an irrigation policy, the five water authorities were successful in developing a common regulation at the regional level. Other regional actors were also invited by the water authorities. However, the LTO and NMO, the respective representatives of individual farmers and local nature conservation organizations, were not able to effectively participate. The development process of the irrigation policy constitutes a typical example of upscaling where the irrigation issue was scaled up from the local level to the regional level. Additionally, the EU level creates pressure on the water boards to work together. According to the WFD, all these water boards are in the Rhine-East and they need to collaborate on water planning and management, of which irrigation management constitutes a significant component. The fact that the water boards have a history of maintaining good relations creates additional impetus regarding the enforcement of the irrigation policy.

Two national measures in the Netherlands also have regional and local implications for agriculture. A national agreement foresees that water from the rivers can be transferred from other regions. However, the water boards differ in terms of the practicality and costs of transferring water from other regions. For instance, WGS, the regional water authority of Salland, has abundant water that could be transferred to its region, whereas it is more difficult for Vechtstromen to bring water from other regions. The positions regarding the responsibility of water authorities in providing water to farmers differs between the water boards that have sloping areas and those have mostly flat areas. The displacement chain is also implemented at the national level to balance the water supply and demand in cases of extreme water shortages by transferring water from abundant areas and sectors to scarce areas and sectors. The decisions are taken by a national committee that involves representatives from all provinces and the functioning of the whole system is controlled by the Rijkswaterstaat, the Dutch national water authority.

In the **Vilaine** catchment, there is little evidence on the interaction of national and subnational levels regarding agricultural measures on drought resilience. However, the implementation of CAP influences the agricultural production and water consumptions at the farm level. As explained in Sect. 10.3, measures that result from the CAP are fragmented from the other water- and agriculture-related measures, since there is no regional agricultural water planning and management. Regional water management plans however address drought and water scarcity through enforcing limitations and bans on irrigation, especially during the summer and in the sensitive areas.

The different actors and levels in the **Somerset** region build up a strong relationship and culture of cooperation. But the vulnerability of such relationships can be seen after the flood events in 2013/2014. Discussions about the nature of the floods and possible solutions eroded the confidence between the different stakeholders and actors pulled out of the circle. On the community level, especially relevant for the implementation of agri-environmental measures, the Farming and Wildlife Advisory Group was seen. They are seen as an in-between agent between environmental groups and farmers. But they are dependent on project funding and therefore exists the risk of reduced activities in case of a reduction of funding. Positive interlinkages between farmers and Internal Drainage Boards (IDB) exist. IDBs are risk management authorities responsible for maintaining rivers, drainage channels, pumping stations, etc. IDB was noticed as a group that is very responsive to farmers' needs. A good relationship was also built up between the local National Farmers Union (NFU) and the farming community. Room for improvement is seen for the relationship between farmers and statutory bodies. Their relationship is not characterized by an open discussion culture but rather by punishment.

A diversity of governance levels and their interactions is observed in the North-west Europe region. In Flanders and Somerset, subnational stakeholders such as farmer organizations and regional water authorities closely collaborate. In other cases, the relevant EU policies explicitly drive cross-level interactions, while there is little evidence that these directly influence the adaptation of the agricultural sector to drought and water scarcity conditions. In the Eifel-Rur, Salland and Twente cases, the WFD requirements create multilevel interactions, such as the collaboration among the regional water authorities that are in the same river basin as well as the participation of local stakeholders in water planning and management decisions, whereas in the Vilaine case, the CAP influences water consumptions at the farm level, although the CAP measures are fragmented from the other water- and agriculture-related measures.

10.5   Public and Political Awareness on Agricultural Effects of Drought

Since drought is an emerging policy issue in most of the North-west European countries, the awareness within the public and political spheres is of crucial importance regarding both the effective implementation of existing measures as well as the formulation of additional measures in the near future.

On the whole, in **Flanders** drought is not yet an issue compared to the perception of flooding impacts for the region. The awareness for water scarcity and drought problems is very low for some stakeholders. Awareness of problems among farmers is growing, but they still want to use groundwater resources today and do not integrate the perception of future generations in their actions. Instruments for awareness raising of different stakeholders are not clearly defined, either. The problem is mainly framed as an agricultural issue, so that the focus of the existing discussion on droughts is more on agriculture and economic development, e.g. compared to consequences on nature areas. The aim of the VMM DROP pilot case was the development and use of indicators for the monitoring and reporting of the drought situation and the modelling of drought impacts using this measurement network. The activities also focused on relevant indicators for the agricultural sector. Furthermore, a coordination platform for drought was initiated that brings together different governmental agencies and organizations involved in water management and agriculture, such as the Flemish agricultural department, regional and national water managers, the provinces and municipalities.

Awareness on drought issues is low in **Eifel-Rur**. Concrete drought-related measures are not taken up for agriculture. It seems a hard task to convince actors of the benefits of working on drought preparedness, as drought events in the region occur very far in between. But farmers start to notice the problem of negative water balance during dry summers. Especially, because the farmers shift to specialized crops requiring occasional irrigation which further influences the water demand. Addressing droughts could be done—at least partially—within other, broader initiatives. For instance, the topic of water scarcity (structural) rather than drought (short term and very far in between) could prove a better banner under which to propose actions that increase system resilience. Climate change can play a role in these debates, as it is predicted that climate change will increase resource use conflicts.

In the **Salland** region, many actors adopt a supply-oriented approach to water. Thus, the major goal is providing the right amount of water with the right quality for all water users. For the water authorities, these users mainly include the agricultural users. The focus on supply has been shifted, since other interests became important in the past few decades as it was realized that the amount of freshwater is limited and climate change is exacerbating this issue. For the agricultural sector, flood protection is still the major goal, whereas drought is seen as a relatively new issue. Although the dry sandy soils of the eastern Netherlands are prone to drought, it is difficult to create a broad awareness of the general public, and especially the farmers, who are concerned more about wet fields and high groundwater levels than the dry fields. There is also a knowledge gap on the appropriate groundwater levels for both agriculture and nature and how they affect one another at various scales.

In the case of **Vilaine**, drought and water scarcity are not seen as urgent issues for the agricultural sector. This can be mainly attributed to the historically favourable situation in the region and the crisis management approach. Thus, the awareness and understanding of both the water users and water managers on the potential impacts of climate change remains weak. The connection between surface water and groundwater resources are not well known, which is partly due to the fact that water withdrawals are not monitored.

In **Somerset**, there is a coherent agreement between the different actors that droughts are already problematic for the region and that these problems will increase in the future. It seems that farmers are quite aware of the problem, but the level of recognition is seen as lower for agriculture and higher for other sectors such as nature. Not the full range of possible mitigation measures on farms are taken up, e.g. rainwater tanks are only installed if there is a subsidy available and is motivated more by saving money. Furthermore, the perspectives on drought are driven by sectoral views, e.g. by agriculture and nature. With further programmes the problem definition could become more coherent. But already before and especially after the flooding in 2013/2014 droughts are seen as a secondary problem to flooding in the region. Opportunities for co-benefits between flood protection and drought protection via the launched action programmes are possible but are not central in the design of activities regarding flood protection. In political discussions there is a reluctance to point out co-benefits between drought and flood measures.

Drought and water scarcity is also a low profile issue in the **Twente** region of Vechtstromen, limiting the financial and political support that could be given for preventive measures. This is mainly due to the historically developed artificial system for managing water levels, which is seen in the west of the country as a sufficient drought management measure, and the conflicting priority for preventing saltwater intrusion by flushing this artificial system. Thus, the eastern part of the country that has areas with high and sandy soils, which depend on rainwater and groundwater, does not receive political priority in terms of drought and water scarcity. The new Delta Programme, which recognizes the climate change and its impacts, is expected to contribute to an improved political awareness for drought in all areas of the country.

Since drought is not yet perceived as an urgent issue in the six case study regions, the overall awareness regarding the agricultural impacts of drought is assessed to be low, especially in the Eifel-Rur, Vilaine, Salland and Twente cases. This situation is also closely related to the historical context, which involves water problems that result from too much water, i.e. floods, rather than drought and water scarcity. Nevertheless, there is an increasing trend in terms of the awareness of the actors that are in the mostly affected areas. These actors mainly include the individual farmers and farmers' organizations, as observed in the cases of Flanders and Somerset.

10.6 Conclusions

This chapter presented a diagnosis on the current state of the governance of drought adaptation in North-west Europe from a cross-cutting perspective on agriculture. The findings indicate both several common regional level implications, as well as issue-specific observations for the local contexts.

Despite being a key user of freshwater resources and often sensitive to water availability during drought periods, the agricultural sector receives a lower priority compared to other water user sectors, which mainly include energy production and drinking water. Pressures to monitor water withdrawals and enforce water withdrawal limitations in agriculture can be expected to intensify with increasing demands not only from the agricultural sector, but also by cross-sectoral impacts of other water user sectors. This competitive disadvantage can lead to water efficiency improvements in the sector through, for instance the dissemination of water saving and water reuse technologies. Additionally, the demand for crop insurances could increase as well as the need for further insurance products such as cooperative private–public insurance products.

Since agricultural production constitutes an economic sector in many areas of North-west Europe, multiple governance levels, ranging from the local to the EU level, interact regarding the associated water problems, including water scarcity and drought. However, there is a clustering of the local and subnational levels, which predominantly shape the implementation of water- and agriculture-related measures, and the EU-level measures that result from the corresponding water and agricultural policies, i.e. the WFD and the CAP.

Although agriculture is a very vulnerable sector regarding the impacts of drought, the essential public and political awareness on drought is limited. The historical context of the North-west Europe region, which is dominated by events of too much water rather than water scarcity and drought, plays a significant role in the relatively low awareness regarding the current and future impacts of drought. In the areas that are already affected by drought, especially the regional authorities, such as environmental agencies and water authorities, and the farmers and their organizations carry the greatest potential in terms of improving drought awareness in both public and political spheres.

References

Bezirksregierung Köln (2013) Strukturdaten 2013 für den Regierungsbezirk Köln. Köln. http://www.bezregkoeln.nrw.de/brk_internet/leistungen/abteilung03/32/regionalplanung/regionalmonitoring/strukturdaten_2013.pdf. Accessed 14 Dec 2015

Bouraoui F, Grizzetti B, Adelsköld G, Behrendt H, De Miguel I, Silgram M, Zaloudik J (2009) Basin characteristics and nutrient losses. The EUROHARP catchment network perspective. J Environ Monit 11(3):515–525

EEA: European Environment Agency (2009) Water resources across Europe—confronting water scarcity and drought. EEA Report No 2/2009, Copenhagen

Geng G, Wu J, Wang Q, Lei T, He B, Li X, Mo X, Luo H, Zhou H, Liu D (2015) Agricultural drought hazard analysis during 1980–2008. A global perspective. Int J Climatol. doi:10.1002/joc.4356

Landwirtschaftskammer Nordrhein-Westfalen (2012) Zahlen zur Landwirtschaft in Nordrhein-Westfalen 2012. Bonn. http://www.landwirtschaftskammer.de/wir/pdf/zahlen-landwirtschaft.pdf. Accessed 14 Dec 2015

Platteau J, van Bogaert T (eds) (2014) Agriculture horticulture 2013 flanders, Department of Agriculture and Fisheries Brussels. http://lv.vlaanderen.be/nl/voorlichting-info/publicaties/studies/report-summaries/agriculture-horticulture-2013-flanders#sthash.0hQ4ws0e.dpuf. Accessed 14 Dec 2015

Sepulcre-Canto G, Horion SMAF, Singleton A, Carrao H, Vogt J (2012) Development of a Combined Drought Indicator to detect agricultural drought in Europe. Nat Hazards Earth Syst Sci 12(11):3519–3531

UN Department of Economic and Social Affairs Division of Sustainable Development (2004): CSD-12/13 (2004–2005) Freshwater Country Profile Belgium, Section G

Chapter 11
Cross-cutting Perspective Freshwater

**Carina Furusho, Rodrigo Vidaurre, Isabelle La Jeunesse
and Maria-Helena Ramos**

11.1   Introduction

One singularity of northwestern Europe (NWE) is that severe droughts are rare events in the region and water scarcity has hardly been experienced in its history. The DROP pilot sites are not exceptions to this context. Although the lack of a drought history in wet areas can explain why drought and water scarcity are not necessarily the focus of (if ever considered in) river basin management plans, it must be noted that freshwater availability for drinking water provision remains a priority stake in both quantitative and qualitative aspects. Providing a reliable and safe supply of drinking water may thus be a leading entryway to the development of drought risk awareness and drought adaptation measures in a river basin. When such essential resource is threatened and the competition for water among users increases, there is a good chance that reflections and changes will be triggered.

Water use conflicts and drinking water supply threats may arise due to increased water demand, but also due to decreased water availability. The later may occur because of natural climate variability, i.e., drier years than average, or as the result of the impact of climate change on local water resources. Climate change awareness is then an important asset to manage water availability. Where climate change awareness is low and adaptation measures are basically inexistent, social and political responses to drought adaptation may be slow and inefficient. However, even in those cases where climate change awareness is still low in general society,

C. Furusho (✉) · M.-H. Ramos
IRSTEA, UR HBAN, 1 rue Pierre-Gilles de Gennes CS 10030, 92761 Antony, France
e-mail: carina.furusho@irstea.fr

R. Vidaurre
Ecologic Institute, Pfalzburger Str. 43/44, 10717 Berlin, Germany

I. La Jeunesse
Université de Tours, UMR CNRS 7324 Citeres, 33, allée Ferdinand de Lesseps,
B.P. 60449, 37204 Tours cedex 3, France

© The Author(s) 2016
H. Bressers et al. (eds.), *Governance for Drought Resilience*,
DOI 10.1007/978-3-319-29671-5_11

Fig. 11.1 The Drézet drinking water plant, in the Vilaine catchment. *Photo* Carina Furusho, 16/09/2013

water authorities and other stakeholders are conscious that water demand tends to intensify with population and economic growth, rendering water scarcity conceivable and even foreseeable.

Freshwater availability for drinking water supply is therefore an issue that can motivate the introduction of drought and water scarcity risks into the political and public agenda, even in "drought-scarce" regions. This chapter highlights the links between drought governance and the vulnerability of freshwater for drinking water supply, with a focus on drought adaptation. The main issues presented here are illustrated with how freshwater issues are managed in the DROP project cases with a particular focus on the two "freshwater reservoir" pilot sites: the Arzal dam in Brittany France (see Chap. 6) and the Eifel-Rur in Germany (see Chap. 4). Those two cases deal with reservoir management not only for drinking water supply (Fig. 11.1) but also for other uses, with various priority sets.

11.2   Drinking Water Scarcity Risks

11.2.1   Relashionship between Water Quality and Water Quantity for Freshwater Uses

During drought episodes, water quality in lakes and reservoirs generally shows deterioration due to less dilution, particularly for nutrients and salinity (Mosley 2014). The increase in salinity observed in most lakes and reservoirs during droughts has been often attributed to reduced flushing/outflows and evapoconcentration, rising concentrations of components due to evaporation (Mayer et al. 2010; Mosley et al. 2012; Burt et al. 2014).

Although the IPCC fourth assessment reports that an increase in average temperatures of several degrees as a result of climate change will lead to an increase in average global precipitation over the course of the twenty-first century, this amount does not necessarily relate to an increase in the amount of drinking water available. A decline in water quality can result from the increase in runoff and precipitation. While the water will carry higher levels of nutrients, it will also contain more pathogens and pollutants. These contaminants were originally stored in soils and in some groundwater reservoirs but the increase in precipitation will flush them out in the river (IPCC 2007).

Similarly, when drought conditions persist and groundwater reserves are depleted, the residual water that remains is often of inferior quality. This is a result of the leakage of saline or contaminated water from the land surface, the confining layers, or the adjacent water bodies that have highly concentrated quantities of contaminants. This occurs because decreased precipitation and runoff results in a concentration of pollution in the water, which leads to an increased load of microbes in waterways and drinking water reservoirs (IPCC 2007).

Water quantity and water quality are thus intrinsically related either in the case of single or multipurpose reservoirs. Their dynamics can be complex, with implications on reservoir operation and control. In the case of the freshwater reservoir of the Vilaine catchment in Brittany, France (Chap. 6), the operation of the locks of the Arzal dam, an estuarine dam in the Atlantic Ocean, is one of the main aspects that influence the quality of the water in the reservoir. The increase in salinity is aggravated by the salt intrusions from the estuary through the opening/closing of the boat lock of the Arzal Dam. The water quality upstream the Arzal Dam is essential to the Drezet-Férel water plant, which provides more than 15 million m^3 of clean drinking water per year to the surrounding population. Salt intrusion deteriorates water quality and provokes the use of siphons that pump water out of the reservoir, back to the ocean. Freshwater is often lost, unavailable for drinking water supply. Integrated quality–quantity management is crucial, notably during summer, as this is the period with highest water consumption, increased number of lock openings for touristic boats, but also the low flow period of the Vilaine River, which is the main inflow of surface water to the reservoir.

In the case of the freshwater reservoir in Eifel-Rur managed by the WVER water board, Germany (Chap. 4), it is mainly the increase in water temperature during drought and low flow periods that can be a serious constraint for drinking water supply. Water must be less than 10 °C to comply with the strict requirements of the German Drinking Water Ordinance. Drinking water regulation limits can be exceeded for a period of 30 days, but only under certain critical conditions. Warmer temperatures not only increase the rate of evaporation of water from the surface of the reservoir into the atmosphere (loss of water quantity), but may also affect water quality, interacting with the amount of organic material in the water, the concentration of pollutants. When the water is warmer, its ability to hold oxygen decreases. The health of a water body is dependent upon its ability to effectively self-purify through biodegradation, which is hindered when there is a reduced amount of dissolved oxygen. Consequently, when precipitation events occur, the contaminants are flushed into waterways and drinking reservoirs, leading to significant health implications.

Although freshwater is the main issue of the two pilot cases mentioned above, other DROP cases also face challenges concerning drinking water provision due to the risk of droughts and water scarcity. For instance, due to its hydrographical situation, water quality in Flanders is subjected to strong impacts on their water volumes and quality caused by upstream countries (see Chap. 7). When interviewed about drought and water scarcity, stakeholders insisted that a rigorous transnational agreement on water volumes and quality crossing the border is essential to avoid political tensions. Another example is the case study in the United Kingdom (see Chap. 5). In order to improve service and quality standards related to drinking water, water companies have been privatized since 1989, in order to increase investment in water and wastewater infrastructure (Water UK 2015). Finally, for both pilot cases in the Netherlands (see Chaps. 8 and 9), water quality has been mentioned as an issue that has been well regulated by successive programs, among which the most recent one is the Delta Decision Freshwater in 2015. Ensuring sufficient freshwater for all water uses, including the environmental ("nature") perspective, is in principle a public task in the Netherlands.

11.2.2 The Diversity of Water Consumption Monitoring Situations

Besides intensifying the challenge of maintaining freshwater quality and quantity for drinking water provision, drought and water scarcity planning also requires better monitoring systems of withdraws to manage water flow and freshwater availability. Monitoring water use, particularly for groundwater, is an issue that is treated differently in each site studied within the DROP project. For instance, in the Vilaine pilot, we observe that only withdrawals related to drinking water are systematically monitored. The knowledge on the water extractions for other uses (industrial, irrigation, and livestock) is much more fragmented because it is not relayed to the water administration, even though it is a legal obligation.

In Groot Salland, in the Netherlands, the water boards ask each farmer once a year to inform about their water extraction levels, although they have concluded that this information is not accurate enough to manage water flows and groundwater levels. Farmers rarely admit having exceeded withdrawal limits. To face current monitoring challenges there are plans to introduce flow meters to monitor water withdrawals at the field. Stakeholders of different water sectors in Flanders also believe that providing drought-risk-related data and good risk communication are essential to incorporate drought risks into their risk management practices in their business. The situation is quite different in Eifel-Rur, where stakeholders indicated that systematic water metering is still not under discussion. The insufficient data collection for flow management could be related to the lack of updated legal requirements.

11.3 Different Priority Settings and Potential Tensions

The fact that floods and droughts are semantically opposites does not mean that any flood control measure is necessarily hindering drought risk management. Conversely, they should not be dealt with separately. People have been fighting against flood risk in all these regions for a longtime, and a dynamic synergy has built among stakeholders. It was clear that stakeholders got used to work together and discuss water-related problems. In that sense, flood risk governance has contributed to bridging connections between stakeholders that can potentially enhance drought governance. However, in terms of synergies, it will also become increasingly important to ensure that the policy measures, and concrete strategies and instruments designed to deal with flooding for each region, are not counteracting any policy developments made for drought and water scarcity.

Drinking water production and flood protection are the main objectives guiding the dam management of both water boards in Vilaine and Eifel-Rur. However, there is a subtle difference that can be noticed when discussing with stakeholders in the way these two priorities are handled by the water boards, reflecting some divergence in perceptions, flexibility, and regulation context between the two cases.

The management rules of the Arzal dam, appended to the Water Management Plan (Schéma d'Aménagement et de gestion des eaux SAGE, see Chap. 6), reflect the hierarchy of objectives to be achieved. Drinking water provision is the first priority and it is widely accepted by all stakeholders interviewed in the Vilaine governance assessment meetings.

In Eifel-Rur the obligation of the water board to provide a well-established level of protection against floods seems to overcome the guarantee of continuous drinking water production. In this context, adapting the dam management rules to prevent water scarcity, even when there is a clear deficit of precipitations (reservoir recharge), is quite troublesome. For this reason, achieving all the high water quality standards demanded by German regulation can be very complicated in drought situations. The strategy for flood prevention in Eifel-Rur implies that the water level in the reservoir must be kept sufficiently low during the winter until the spring to ensure enough storage capacity in case of exceptional flood events which may be associated with intense rainfall or snow melt. However, if there is not enough precipitation or snow melt during the spring period, when water is collected, there is not enough water to meet all the quality conditions for drinking water providers (e.g., water temperature below 10 °C and oxygen above 4 mg/l). It is a lengthy process to change the flood protection rule to adapt the reservoir level for drinking water purposes in cases when precipitations arrive earlier than expected. The water board first needs to prove, based on data analysis, that the proposed changes would not compromise safety-concerning flood risks. In this sense, the requirement for evidence (based on simulations using historical data) can slow down the implementation of adaptation measures: the legal aspects bring with them a reluctance to take responsibility to adjust management rules without clear science-based evidence.

Water use ranking in case of drought and water scarcity is a subject that has not been highlighted by stakeholders in the DROP project interviews in Somerset, but drinking water and environment tend to get priority all over England with different expressions in regions according to the Water Act 2003. The priorities established by the Dutch national "*verdringingsreeks*" (displacement chain) in case of serious freshwater shortage are not the same as in France and Germany. Preventing irreparable damage to the water system, the soil (e.g., peat layers) or nature is the first priority of the chain. Drinking water and energy production come as second in line, followed by high-value agricultural and industrial production processes and last by the interests of shipping, general agriculture, nature with resilience, industry, recreation, and fishery.

Surprisingly, there is no "hierarchy" or prioritization of different water uses/demand if a situation of water scarcity occurs in Flanders. The VMM water board, which is developing physical drought indicators provided by modeling assessment tools for the monitoring and reporting of the drought situation, is now getting this issue on the agenda. The fact that drinking water companies set lower prices for large-volume consumers, as some industries, does not contribute to regulating demand and is not coherent with the general aims of the water board, particularly in the perspective of preserving environmental flow.

In Vechtstromen, the second DROP case study in the Netherlands (see Chap. 9), increasing extractions for irrigation and drinking water threatens the groundwater-sensitive areas. As a result, the farmer organizations and drinking water companies are opposed to nature conservation organizations. Province and Vitens (the local drinking water producer) are looking for ways to protect drinking water resources by combining nature and drinking water protection through the involvement of water boards and farmers. Vitens provides financial compensation to the farmers and for nature areas that are affected by its water abstractions.

In Eifel-Rur, the obligation of the water board to provide a well-established level of protection against floods and drinking water supply, with all the responsibilities associated, have resulted in an elaborate and sophisticated set of rules to manage the interaction of reservoirs and water bodies. These legal obligations restrict the possibility of officially incorporating additional risks (e.g., droughts) into the set of priorities which govern the system. Even small changes have to be extremely well founded and well argued, based on technical evidence and modeling of historic data. The overall framework is therefore destined to be rather reactive than proactive, and these reactions tend to take time. The management of secondary objectives or other unconsidered aspects can only be improved if it can be shown that primary objectives are not affected. This means that the adaptation of dam management rules to drought and water scarcity is a lengthy procedure.

11.4   Multilevel and Multiscale Issues and Measures

A comparative analysis of three drinking water provision issues, in the Vilaine, in Eifel-Rur, and in Flanders, can be particularly illuminating, as they present similar problems in very contrasting contexts, different levels, and scales involved as well as a diversity of other factors influencing them. In the case of the **Vilaine**, problems of water quantity related to the Arzal dam reservoir translate into a problem of water quality. As explained in further detail previously in this chapter, in dry periods the low inflow from the Vilaine river and the intrusion of salt water through the lock for sailing boats are increasingly causing water quality bottlenecks for the drinking water plant. The position of the reservoir at the river mouth is downstream the big catchment area affecting the reservoir (of slightly over **10,000 km^2**), which in turn implies **a large scale and a huge number of administrative levels** to be potentially involved in the different possible solutions. This position of the reservoir also means that it is impacted by the water management decisions of many different actors and sectors. A series of sectors (including the traditionally strong agricultural sector) rely on water management, both in terms of water availability and in terms of water drainage, and for decisions affecting the region's water management the different needs have to be aligned between the parties.

Whereas this dependency of drinking water provision on the outcomes of water management measures (such as those derived from the implementation of the WFD) would seem a problematic dependency, in practice the Vilaine catchment water board (IAV) is responsible for both drinking water provision for water companies and for implementing the Water Framework Directive. This means that it is in a privileged position to keep track of issues affecting water quality and react accordingly to possible problems. IAV has recourse to an array of possible solutions to address their water quality issues. For instance, they have the possibility of implementing measures throughout the catchment in order to avoid excessive water level drop in summer months by adding small dams along the stream and tributaries. However, these options seem less attractive than improving the lock system to decrease saline water intrusion, which is the solution that the IAV is currently evaluating using a prototype (within the DROP project framework). The solution addresses an existing inefficiency and does so at one point which is under the management control of the water board. Decentralized options may require the cooperation of other stakeholders and continuous efforts over time and therefore seem more complex to implement efficiently.

In the case of the **Eifel-Rur** region, dry years also create water quality problems in one reservoir, but these problems are of another kind, as they are related to issues of eutrophication. Dry years thus mean that the quality and temperature of the water provided by this particular reservoir can be compromised; creating issues for the drinking water company supplied by the water board. The issue is **very limited in scale, as the affected reservoir is upstream within the watershed**. The reservoir's catchment area is mountainous, mainly forested (i.e., not much agriculture), has hardly any population, and with a size in the order of **a few hundred km^2**. This implies on the one hand that there are not many actors to be dealt with, whose interests would have to be aligned in possible measures. On the other that there are few control structures affecting springtime water availability which could be managed to improve water availability. Indeed, one possible solution is to adjust management plans so as to allow for more "winter water" to be kept in the reservoir under dry hydrological conditions; the increase in water quantity would help to avoid the decrease in water quality and its temperature rise.

The approach chosen by WVER—to adjust operating rules to be better prepared for dry years—is thus an issue requiring interaction with few stakeholders. The problem is fundamentally one of legal responsibility (how to increase "winter water" in the reservoir without affecting the water board's other legal requirements such as flood protection; this issue could potentially be related to expensive litigation), so discussions are directly with the relevant authority. Since the required agreement involves only authorities and the water board, the scale of the reservoirs management in Eifel-Rur is quite limited compared to the Arzal Dam management.

Flanders relies on a mix of groundwater and surface water for its drinking water provision, and summer low flows in the large transboundary rivers that cross the country are accompanied with water quality issues. In recognition of this problem (which is not new), water companies have infrastructure which allows the retention of higher quality "spring" river water for use over the summer months. However, longer dry periods mean that this buffering capacity no longer seems sufficient, and both authorities and drinking water providers admit the necessity of increasing the volumes retained—which means building additional retention infrastructure. The water quality of the rivers that flow through Flanders is beyond the control of the region or even of Belgium, as these are large international river basins (Meuse: **34,548 km^2**; Scheldt: **21,863 km^2**) covering a huge geographical scale and levels going up all the way to the international. As an overall conclusion, the drought-related issues affecting drinking water in the northwestern European pilots were not directly a problem of water availability, but of limited water flow generating different water quality consequences. Longer periods of low flow (Vilaine, Flanders) or changed precipitation patterns (Eifel-Rur) affect water quality negatively, to the point that drinking water companies see the need for (sometimes expensive) action. In all three areas, and in spite of the largely different scales, the planned responses were related to infrastructure: improving infrastructure by eliminating existing inefficiencies (Vilaine), increasing the capacity of infrastructure (more reservoir capacity in Flanders), or adjusting operational rules of infrastructure.

11.4.1 Coordination Above Local Level for Increased Resilience

When it comes to drinking water supply, the case study areas exemplify a broader trend of increasing spatial water connectivity between neighboring water service

Fig. 11.2 Drinking water provision network of the Vilaine catchment and connections. Map displayed at the Drézet water plant. *Photo* Isabelle La Jeunesse, 16/09/2013

provision systems. This development is usually the result of contingency planning, and sometimes the result of legal requirements for contingency preparedness. This increased connectivity does not target exclusively or even primarily the risk to water provision due to droughts (they address many different risks that may interrupt water service provision), but it does enhance preparedness for drought episodes. The solutions emerging in the northwest of Europe illustrated by case studies analysis also reflect this perception of a scale expansion in connectivity to improve the robustness of drinking water systems.

In the Vilaine, the first phase of the interconnection between drinking water networks has been implemented (Fig. 11.2) and will be expanded according to the SAGE. In Eifel-Rur, the technical solutions to improve the water system robustness and develop backup solutions in case of extreme water scarcity were mentioned by the drinking water producer and also by the hydroelectricity power plant manager. There is the possibility to connect their system to the Mosel River, for instance. The same trend has been noted in Flanders, where drinking water companies acknowledge the need for additional buffering capacity by enlarging the infrastructure interconnectivity among catchments.

Drinking water companies can be public-owned, privatized, or public-owned private companies. In the Vilaine and in the Eifel-Rur, drinking water provision is under the responsibility of public institutes (IAV and WVER water boards). In the

Netherlands, water supply companies are publicly owned private companies, with often dozens of municipalities and provinces as owners. They are submitted to the national "drinking water regulation" determining the maximum return for invested capital, therefore regulating the price of tap water. The companies have no pressure to maximize prices and instead have a sort of corporate pride in delivering good quality water for a modest price.

The UK has privatized drinking water companies. They are responsible for the abstraction of water from rivers and streams and aquifers for drinking water supply, but they also have a range of roles and responsibilities in environment conservation and drought and climate change adaptation planning. Their company borders do not necessarily map onto watersheds. Even in the context of this particular setting, the full range of administrative levels and scales are involved in drought management and water scarcity for drinking water in the Somerset region. However, this setting also creates some cross-boundary issues that span drinking water supply, environmental flow, and agricultural water use. The water companies have a drought plan that covers drinking water supply (in balance with other environmental factors like flow), but the Environmental Agency has another drought plan that includes both water supply and irrigation issues covering a region rather than just a water company.

11.4.2   Larger Scales for Long-Term Strategies

Moving up to the regional-level implication in drinking water supply, in Eifel-Rur, the district level focuses in long-term development of regional water management. In Vilaine, the regional coherence in terms of water planning is ensured by the SAGE (Schéma d'Aménagement et de Gestion de l'Eau). The sustainability and the quality of the drinking water resource is the major issue that framed the SAGE Vilaine and the debate between all actors involved. Similarly in Flanders, the regions are the ones responsible for water policy, including drinking water quality.

The economic aspects of drinking water provision (i.e., the establishment of maximum prices and the approval of price increases) are often managed at the national level. That is the case with the Federal Government in Belgium and also in the UK, where the OFWAT (the Water Services Regulation Authority) is the financial and economic regulator of the water and sewerage sectors. They have a duty to set the price, investment, and services standards. In France, the legislation designates that "drinking water pays for drinking water", imposing an independent budget of drinking water supply and other water management sectors. The price of water is also fixed and indexed to the cost of its management.

Drinking water supply is also dealt with in transnational economical arrangements, as the Eifel-Rur drinking water producer sells water to Belgium and the Netherlands. In Flanders, a key instrument that seems to be missing is the transnational agreement of flows over borders, particularly with France. Drinking water companies complain that the water quality is hard to maintain when flows are reduced, especially during dry summers. The lack of such agreements also delays authorization for the establishment of new drinking water production facilities. Political will to develop a legal framework seems to be lacking, but there is also a problem of leverage of the French government.

Drinking water standards, wastewater discharges, and other issues are also governed by the Water Framework Directive across EU countries. It was noticed that EU environmental policies seem to play an important role to introduce a more holistic and synergistic approach to drinking water supply and the management of the reservoirs. At the same time some stakeholders interviewed in Eifel-Rur expressed criticism of EU regulations, which are seen as "imposed from Brussels". The existence of such a "distant" authority has shown to be beneficial when unpopular measures must be pushed by the water boards, as they can argue that they have no choice but to comply with EU directives.

11.5 Awareness and the Public and Political Agenda

The interview campaigns held within the DROP project highlighted that the broad public is in general unaware of the risks and challenges water providers are facing due to drought. Users are accustomed to a high quality of service 24 h a day, 7 days a week; service interruption is seen as someone not having done his homework, rather than a possibility that can arise as a result of different natural risks to service provision. In addition, stakeholders highlighted that the broad public is typically unaware of the sources of their drinking water. In the Eifel-Rur region, for instance, the overall public perception is that the reservoirs provide other more visible services than drinking water, such as flood protection or opportunities for sailing and tourism attraction. This lack of awareness is a drawback when trying to communicate drought risks to the broader public (La Jeunesse et al. 2015).

Communication on droughts faces additional challenges in these flood-prone regions. These highly visible impacting events convey to the broader public the idea that a certain region's problems are related to dealing with too much water, and not too little of it, as far as reservoirs are managed for protection against floods and also sustain stream flows during low-flows periods. Conveying the concept that flood risk does not imply an absence of drought risk is a communicational challenge.

Awareness of the topic among stakeholder groups seems not much higher than that of the broader public. Stakeholders, in general, do not consider drought and water scarcity issue as urgent from their perspective, and there is a lot of interest in keeping up business as usual or even in expanding water uses. The exceptions to this rule are the drinking water providers themselves—some proof is given by the fact that the water boards IAV and WVER are part of the DROP project, and in Flanders drinking water providers also counted this issue as on their agenda. Beyond drinking water providers, some environmental authorities considered were showing interest in the issue, fundamentally due to the environmental problems that could derive of the low flows. Somewhat surprisingly, environmental NGOs in the

Vilaine region, Eifel-Rur, and Flanders saw the topic as an issue but not significant enough to consider it one of their priorities.

It is probably for this reason that drought is not very present on the political agendas of the analyzed regions: since stakeholders groups as yet mostly are disengaged with the topic, there is no pressure by the electorate or by interest groups on the political or administrative levels to support this topic. In addition, issues of water use and expansion of water use often involve strong economic interests. Stakeholders express that it can be very hard to argue against economic uses of water. In the Eifel-Rur region the paper industry and farmers have significant political influence, also related to the amount of jobs they create in the region. A similar situation was observed for farmers in Flanders and Somerset (UK). With the current political agenda very much pro-growth, it would seem that there is not much potential for the uptake of an issue which stakeholders reject due to the possible impacts on business opportunities.

11.6 Conclusion: Diagnosis and Scenarios

Currently, drought management practices in NWE are largely based on crisis management. The effectiveness of these practices is questionable because they are reactive, dealing only with the impacts of drought rather than tackling the causes of the vulnerabilities. This does not promote the anticipation of adaptation strategy development while measures can require time to be operational. Proactive management has generally been implemented in case studies following drastic droughts (Dennis 2013; Krysanova et al. 2008). The consequences of disasters can create sufficient public and institutional willpower to lead authorities and stakeholders to design and implement proactive approaches to mitigate impacts of future drought episodes.

In the case of the Northwest European region, there is still a visible inertia to start moving toward the development of adaptation measures to improve drinking water supply systems' robustness. This inertia seems to be mostly due to the lack of severe drought and water scarcity episodes in the collective memory that motivate other regions to mobilize stakeholders of all levels to tackle these problems when they are really experienced.

Even in these cases where climate change awareness is still quite low and where drought and water scarcity have hardly been experienced, the essentiality of drinking water supply and freshwater availability may be the leading entryway to the development of drought risk awareness and drought adaptation measures. Most people are aware that fresh water is a limited resource and that water demand is indeed increasing with population growth and economic development. This perception helps them realize that the threat of water scarcity is possible and foreseeable, even if they have not experienced it in the past. That is why the issue of drinking water provision is a key factor to be highlighted to push forward adaptation measures to prevent drought and water scarcity.

One important step toward this objective is the implementation of better monitoring systems of water withdraws to manage water flow and freshwater availability, as it has been highlighted by the analysis of the DROP pilot sites. In fact, besides monitoring water withdraw, all the data that can contribute to a better understanding of the water cycle is worth being collected to provide the basis for science and best practices in hydrology, water supply systems, geomorphology, drainage network, and land use management. An enhanced knowledge of drought impacts and of hydrologic patterns contributes to achieving greater effectiveness of adaptation measures and target management efforts.

The well-developed flood risk governance in pilot cases seems to have contributed to creating synergies among local stakeholders that can participate in building integrated water-related risks (including droughts) governance together. Future actions that could enhance drought resilience include the following strategies (selected from the study of Dennis 2013):

- New sources of water from outside the region are pursued to meet demands (drinking water supply systems interconnectivity).
- Residents collect gray water for outdoor use.
- Cities utilize policy instruments (like financial incentives) to reduce water use.
- Water quality regulations are precautionary and protect against new and potentially harmful pollutants.
- Natural areas along streams are restored and protected for fish and wildlife.
- Safe yield is a central guiding principal in water management.

The evolution of regulations and policy instruments depends greatly on changes of the political agenda in the region, the main topic was discussed in the previous Sect. 11.5 of this chapter. The first point actually concerns measures that have already been identified and even started to be implemented by water managers in NWE. However, the actions that require a paradigm shift to a most systemic strategy including water demand control remain out of the agenda and could greatly improve the resilience of the region to drought and water scarcity rising risk.

References

Burt TP, Worrall F, Howden NJK, Anderson MG (2014) Shifts in discharge-concentration relationships as a small catchment recovers from severe drought. Hydrol Process 29(4):498–507. doi:10.1002/hyp.10169 Accessed 16 Dec 2015

Dennis L (2013) Proactive flood and drought management. A selection of applied strategies and lessons learned from around the United States. http://www.awra.org/webinars/AWRA_report_proactive_flood_drought_final.pdf. Accessed 20 Oct 2015

IPCC (2007) Contribution of working groups I, II and III to the fourth assessment report of the intergovernmental panel on climate change core writing team. In: Pachauri RK, Reisinger A (eds) IPCC. Switzerland, Geneva, p 104

Krysanova V, Buiteveld H, Haase D, Hattermann FF, van Niekerk K, Roest K, Martinez-Santos P, Schlüter M (2008) Practices and lessons learned in coping with climatic hazards at the river-basinscale. Floods and drought. Ecol Soc 13(2):32. http://www.ecologyandsociety.org/vol13/iss2/art32/. Accessed 21 Oct 2015

La Jeunesse I, Cirelli C, Larrue C, Aubin D, Sellami H, Afifi S, Bellin A, Benabdallah S, Bird DN, Deidda R, Dettori M, Engin G, Herrmann F, Ludwig R, Mabrouk B, Majone B, Paniconi C, Soddu A (2015) Is climate change a threat for water uses in the Mediterranean region? Results Surv Local Scale Sci Total Environ. doi:10.1016/j.scitotenv.2015.04.062

Mayer B, Shanley JB, Bailey SW, Mitchell MJ (2010) Identifying sources of stream water sulfate after a summer drought in the sleepers river watershed (Vermont, USA) using hydrological, chemical, and isotopic techniques. Appl Geochem 25(5):747–754

Mosley LM (2014) Drought impacts on the water quality of freshwater systems; review and integration. Earth-Science Reviews, vol 140. Jan 2015, pp 203–214. ISSN 0012-8252 http://dx.doi.org/10.1016/j.earscirev.2014.11.010

Mosley LM, Zammit B, Leyden E, Heneker TM, Hipsey MR, Skinner D, Aldridge KT (2012) The impact of extreme low flows on the water quality of the lower Murray River and Lakes (South Australia). Water Resour Manage 26:3923–3946

Water UK (2015) Price reviews. http://www.water.org.uk/price-reviews. Accessed 12 Dec 2015

Chapter 12
Cross-cutting Perspective on Nature

Hans Bressers and Ulf Stein

12.1 Introduction

In this chapter, we delve into the nature perspective in order to supply cross-cutting insights to the interlinkages between nature and drought. *These interlinkages are not just found in the two regions that have chosen this perspective as the subject of their pilots, but are also to some extent relevant in the other four areas.* For the purposes of this chapter, the term nature is applied as a broad proxy for several nature-related concepts. Most generally, nature here refers to areas designated for nature conservation. This includes nature areas under explicit protection. This concept of nature also extends to natural elements within conservation areas, including river and catchment systems, the diversity of species present and/or threatened, and ecosystems and their ecosystem services. In addition, implicit to this framing of nature is the policy context that extends across nature conservation and land use management measures. This includes sustainable land use policies and practices.

12.2 Drought and Water Scarcity Problems Related to Nature

To understand the relationship between nature and water scarcity and drought problems, we turn our attention to five case studies to illuminate key points. We introduce the areas and their specific relationship to nature and ways in which drought is already impacting natural areas and their ability to cope. The policy

H. Bressers (✉)
University of Twente, P.O Box 217, 7500 AE Enschede, The Netherlands
e-mail: hans.bressers@utwente.nl

U. Stein
Ecologic Institute, Berlin, Germany

© The Author(s) 2016
H. Bressers et al. (eds.), *Governance for Drought Resilience*,
DOI 10.1007/978-3-319-29671-5_12

context is also elaborated to provide insight to the current milieu, including relevant actors and relevant economic and social challenges.

In *Flanders* there are two types of drought sensitive areas. In some parts in the west the horticulture is so dependent on sufficient quantities of good quality water that any present or future disruption in the provision of servicing water is likely to lead to high economic costs. In the higher altitude sandy parts of Flanders that are dependent on groundwater and rain, drought sensitive nature areas no doubt suffer from periods of desiccation, such as in adjacent areas in the Netherlands. However, it is remarkable that how little attention this gets in Flanders.

In the area of *Eifel-Rur*, nature is very important. Large parts of the area are covered by woods with the National Park Eifel spanning approximately 110 km^2. The national park promotes itself as a "wood and water wilderness". Preserving a vulnerable landscape that also attracts tourists is therefore one of the objectives of the water authority, which cooperates with other actors to achieve beyond its legal duties. The national park authority was the only actor to mention drought as an issue threatening this area. Though dryer summers have been the trend across most of the area, until now no severe droughts have struck the region that endangered the role of water for nature. There are limits to increasing water levels in the spring in order to create buffers for dry periods due to the flood protection function of the reservoirs, which require sufficient retention capacity.

The area of *Groot Salland* includes a national park that spans 35 km^2, mainly consisting of wooded hills and slopes and also areas of heather. A large part of it is Natura 2000 area. Several smaller Natura 2000 areas are included in the territory under the water authority. One main concern for the area is the buffer zones, which enable water levels that are adapted to the envisioned land use (higher levels for nature, somewhat lower water levels for agriculture). Moreover, groundwater extractions by drinking water companies and farmers in and directly around the buffer zones have a direct impact on desiccation of nature and pose a threat to the quality requirements of nature management under Natura 2000.

In the *Vilaine* catchment area the centre piece consists of a large wetland area, the Vilaine and Redon Natura 2000 area spaning 100 km^2. Oddly, the main threat of drought for these wetlands is indirect and does not imply desiccation, but flooding. The Vilaine catchment terminates in a big reservoir adjacent to the sea. The water levels of the marshes are controlled by this dam. The water level of the reservoir is adapted to the needs of flood protection and drinking water production and not to the needs of the wetlands. To protect drinking water reserves and prevent salinization, especially during relatively dry periods, the water level of the reservoir is kept up. This has the consequence that during the dry periods the wetlands actually get largely submerged! The building of the dam and reservoir itself also had big impacts on the area: without the reservoir the tide impacted the waterways as far as 60 kilometres inland. After the dam was constructed the accumulation of deposits from the river drove the mussels farming and eels fishery outside the original area. A third and more common problem for the eastern portion of the area stems from conflicting water use needs, where irrigated crop agriculture competes with the needs of the nature areas during dry periods.

As the Visit *Somerset* website (www.visitsomerset.co.uk—June 2015) describes: "The Internationally important Somerset Levels and Moors form a unique patchwork landscape steeped in history and brimming with rare wildlife. Today the area is mostly grassland and arable with willow grown commercially". Not surprisingly, with its rich wildlife, including the largest lowland population of breeding wading birds, the area is specially protected and supports a number of protected areas.

The quote shows both the important natural value of the area, and the fact that it currently is and historically has been used for human purposes. However, the area is vulnerable to both floods and droughts. Climate change is increasing the frequency and severity of both events.

Water shortages due to prolonged droughts can be quite harmful for the wetlands, particularly habitat and peat loss. Agriculture and villages, however, are more afraid of flooding. After some years of severe drought in 2010–2012, a lengthy flood submerged much of the area in the winter of 2013–2014. The peat soils provide various ecosystem services not only for nature itself, but also for carbon storage, food production, and the protection of the historic environment. The impacts of dry periods can create irreversible changes, such as compaction and land subsidence, which not only decreases the buffer capacity of the land, but also increases flood risks down the line.

The water authority of Vechtstromen produced analysis that concentrated on the *Twente* region. In the region, various nature areas suffer from desiccation in dry periods. About 90 % of the creeks run (almost) dry in the summer. Much of this region is dependent on rain and groundwater.

While the area does not have one large designated nature area, it includes many valuable woods, heather fields and wetlands of various sizes in a mostly small-scale landscape. For this reason it was considered an official "National Landscape", and is, despite the termination of this national policy programme, still considered so by all regional actors involved.

Due to climate change the annual water balance gets dryer affecting also groundwater levels. This in turn can influence river discharge while in dry summers creeks are often fed by groundwater. Not only in streams and their valleys, but also on higher ground nature will deteriorate with desiccation. The map below shows both the patchwork of existing nature areas and the areas in dotted green and yellow where even smaller spots of nature that are mingled with agriculture will be developed with a priority for nature.

For most of these nature areas desiccation is a serious challenge, but often nitrogen from air pollution and farming is an even bigger one. The map also shows some main waterways in the rain and sometimes even groundwater fed creek system (Fig. 12.1). The development of drought resilience measures with farmers always has to cope with some distrust that nature development is restricting rather than codeveloping with farming, which can actually be true in some instances.

The overview in this section shows that most areas involved in the DROP project do have *drought sensitive nature areas*, even though their nature and the extent to which they are threatened by droughts vary. While it is very hard to find data on drought impacts on Flemish nature areas, given the characteristics of the area it is

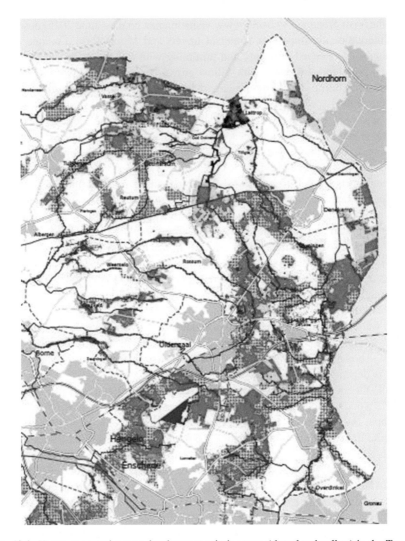

Fig. 12.1 Nature areas and nature development priority areas (*dotted* and *yellow*) in the Twente pilot area of Vechtstromen (*light grey* are cities and towns). *Source* Website Province of Overijssel, page on National Ecological Network

hardly conceivable that there are no impacts. The only regions where nature appears to be rather unthreatened by water scarcity are the two that have water reservoirs. In the Vilaine area, however, this has resulted in the reverse effect, namely that of flooding of wetlands in the dry season. This relates to a second issue, that of the degree of *modification of the water system*. In all regions the water system has been modified in the past, either for purposes of flood protection or agricultural efficiency or both. All of these interventions had side effects on other services of the water system that were often not recognized when the interventions took place.

12.3 Drought Issues and Other Climate Change and Competing Sectors' Risks for Nature

Building on the contextual relationship between drought and natural areas, here we broaden our understanding of drought issues within the context of climate and additional competing sectors to formulate a brief glimpse into the interacting pressures that colour each case study region.

In *Flanders* the attention for the drought consequences of climate change for nature is still in its infancy. If there is emphasis on drought, the consequences for agriculture and other economic interests are the main drivers. Projects in which climate resilience of water systems will be improved are still rare.

The sequence of reservoirs in the Eifel-Rur area serves a flood protection goal. However, the water abstractions are also important and can compete in the future in periods of water scarcity. A minimal flow requirement, that often has environmental purposes, serves mainly to preserve the supply to industries downstream. The water board itself has a council in which various stakeholders are represented including all relevant governments and major water right owners. However, users of smaller quantities such as farmers, fishermen and nature organization have no direct representation. These users and their sectoral representatives, including nature, are invited to roundtable discussions on the initiative of the water board. This proactive network building is highly appreciated by them. In discussions on water abstraction forestry and nature organizations are therefore also involved. Public perception of the function of the reservoirs in flood protection is not very well recognized. Rather, the tourism and sailing functions are seen as main purpose of the reservoirs. In this sense, preventing future drought effects on nature could get some public support. Such measures however will face resistance from water right holders when they intrude therein.

The buffer zones and irrigation around nature areas in *Groot Salland* is a classic example of the competition between nature and agricultural sector interests. While the EU nature policy is influential, involvement of nature conversation NGOs is limited. There have been cuts in the amount and the way the nature NGOs get support from the province, implying that they often cannot send real experts to the roundtables for which they are still invited. In some flat parts of the area it is actually possible to let water in from bigger waterways. This helps farmers, though such external and less pure water is not used directly to restore the water balance in the sensitive nature areas. The pilot project of the water authority in DROP is a waterway that can be used for dual purposes: to get rid of water in the wet season, but also to let water in during droughts. Flood protection and drought resilience are thus combined.

The building of the dam and reservoir in the *Vilaine* catchment area turned a salty marsh into a freshwater area that could more easily be used for farming and drinking water production. The desalination of the marsh was thus a strong intervention in the natural system. Apart from this, flood protection was the main goal since high river waters could not any longer coincide with peak sea levels causing risk of serious flooding. Drinking water production is the priority use for this water. Other goals such as tourism and yachting, agriculture, and nature are acknowledged, but clearly as secondary objectives. Fisheries were hit hard, but recovered by relocation and more recently by creating fish passages and minimal flow requirements. A new anti-salinity lock should enable yachting while preserving as much freshwater as possible and keep the ecological quality high. In this way engineering interventions initiated for flood defence created the need for further interventions to modify ecological side effects.

The clearest example of climate change impacts increasing the frequency of both droughts and flooding in *Somerset*. It is not considered a water-stressed area because on average there is no problem with the supply and demand balance. There can however still be scarcity for nature, especially during times of drought. While droughts in other parts of South England even have been hitting harder, the Somerset nature areas are particularly vulnerable. After the 2013–2014 floods the Somerset Levels and Moors Flood Action Plan replaced a number of drought and flood resilience plans with a combination of measures in which flood protection had an overriding priority. Apart from improved flood defence measures and dredging of waterways, also dredging around the Levels and Moors is included in the plan.

The science is inconclusive on whether this is the most appropriate set of measures, even for flood protection, while it certainly would increase the risks of drought impacts on nature. A catchment approach that also would balance the needs of nature and agriculture is an emerging agenda in the region and country but there is still discussion on the effectiveness of these approaches. Farmers have no incentives now to accept higher water levels at the expense of the usability of their meadows to buffer for dryer periods. Some even muddle with the water system for purposes of irrigation. Another issue is peat extraction for which often old licenses are still valid. Yet another issue is water extraction rights that are not regarded flexible enough for future challenges. Nevertheless the growth of larger conservation NGOs have increased the awareness of the importance of the wetlands for the region.

The Twente region in *Vechtstromen* is in general not a flood-prone area. Nevertheless, heavy rains that are more frequently occurring due to climate change surpass the water drainage capacity in several areas, causing short term flooding ("water on the streets") problems and damages like in the summer of 2013. The main stream in the pilot area is the river Dinkel, a transboundary river that remained mostly unmodified at the Dutch side, while upstream at the German side all meanders were removed and the river was "normalized". This increased the flooding risk of the Dutch stream valley, including both agricultural fields and many nature areas. August 2010 was the wettest month in this area in 100 years, causing the Dinkel to submerge large areas and even parts of the town of Losser. An interesting background is that the German measures were part of a transboundary plan, but while insights in water management changed the Dutch decided to stop its implementation. Nature would have been severely damaged and drought risks increased by the kind of measures proposed in the 70s and 80s. At other places, also in Twente, the water authorities had already started to undo precisely these kinds of measures. Instead they compensated all adjacent farmers by paying them the difference in land value between their land in its natural status and in a well-drained status. Sectoral rivalries regarding nature and drought exist in the pilot area in the form of water extractions in or near nature areas by both drinking water companies and framers. In the nature area of Mander a large drinking water well is exploited.

Fierce discussions between the water authority and the drinking water company were resolved for the time being by a "gentleman's agreement" in 2008 that the extraction would be stopped as soon as alternatives had been found. While working on this, such alternatives still have not been placed into effect, also because there is a structural shortage of water for drinking water production in the Twente region, requiring the drinking water company to even import water from Germany. The water extraction by farmers for irrigation is less of a problem for nature in Twente, since most of the creeks run almost dry in summer anyhow and have permanent extraction bans and pumping groundwater is very expensive. This pressure might rise however in the future as the lifting of the milk quota will make farmers strive for higher productivity. For the rest, similar rivalries than in Groot Salland exist concerning the desired water levels for nature and those for agriculture and the buffer zones that are required for this difference. Promoting the acceptance of generally higher water levels by farmers is one of the drought resilience measures taken.

This overview shows that across the Northwest European area *climate change has a double impact* of intensifying both drought and flood risks, demonstrated by more extreme weather events and periods. While the areas with water reservoirs can cope relatively easily with these impacts also in those cases further climate change effects can disturb that picture in the future. The water scarcity impacts of climate change on nature areas are worsened by sectoral *rivalries* with other water consumers, like drinking water extractions and agriculture, especially when it is irrigated agriculture. In Eifel-Rur also industry and in Vilaine also tourism (yachting) present competing claims on the water availability.

12.4 Multilevel and Multiscale Issues and Nature Measures

Nature protection runs the risk of just leading to isolated patches of natural beauty. To create viable ecosystems their relationships need to be considered. Likewise, also the governance of nature needs to be multilevel with the involvement of all relevant administrative levels.

In *Flanders* the municipalities are handing over responsibilities for small waters to the provinces. As a consequence they get "out of the loop" on these issues. This reinforces that there is very low awareness of drought as a problem and economic interests prevail over environmental ones. Thus if anything, it is the agricultural consequences that get attention at the local level, not the nature consequences.

The consequences of increasing water demand are discussed at the provincial level but not at the local level.

In the *Eifel-Rur* area droughts are not yet seen as a threat to nature. However, strict EU Natura 2000 requirements can impact the measures taken. EU Environmental policies seem to play an important role in introducing a more holistic and synergistic approach to the management of the reservoirs. The water board has a co-responsibility for nature conservation, as the districts have. In addition, local groups like the "Salmon initiative" have started actions that were picked up by the district authorities.

In the area of *Groot Salland*, the issue of the protection and upgrading of Natura 2000 areas is a central concern. Motivated by fears that the Natura 2000 requirements would block any agricultural development in close proximity to nature areas, the Overijssel province has reserved a very large budget for measures in and near the buffer zones, including compensation payments for farmers and if necessary the buying of land to enable farmers a fresh start elsewhere. Because old extraction permits are valid indefinitely, it is difficult to use involuntary measures in the buffer zone. Apart from European-level influence, recent transfer of responsibility from the national level to the provinces plays a role as well. Another multilevel aspect is related to the development of a joint irrigation policy by all the water authorities in the Rhine East subbasin. Whereas in the past, each water authority had its own rules regarding temporary irrigation bans, now this regulatory framework is similar. The nature NGOs found the new policy made too much in a rush and thus, with technical shortcomings. The results were restrictions to new extractions in Natura 2000 areas as well as the buffer zones. However, the existing extractions are left untouched and are not even fully known. The water authorities ask the individual farms yearly to communicate their extraction levels. However, compliance is not monitored.

In France, climate change adaptation, especially with regards to drought, is mainly a national endeavour. In the *Vilaine* area, the Water Plan (SAGE) is the geographically largest in France, and aims to improve the water quality, aquatic environment and the wetlands in the Vilaine catchment. The Water Plan was created by a local water commission with two geographical subcommittees: one for the estuary and one for the wetlands. The latter incorporates elements from the 2008 Natura 2000 plan. For the eastern area, it is relevant that the entire catchment is not regarded anymore a water-sensitive region. As a consequence, the new Water Plan lost its instruments for drought management. The issue of small hillside reservoirs was extensively discussed while some feared that what would be initially accepted for vegetable growth soon would be diverted to irrigate crops like corn and thereby markedly increase the water demand of the area creating future drought problems for the wetlands. A further problem mentioned is that local implementation is difficult since often local politicians are too close to the farmers.

In *Somerset*, the management of the flood crisis illuminated the inherent levels and scales of the problem. Though historic collaboration between the various governance levels of the rather complicated and partially privatized British system, there was an immediate lack of funding on behalf of the higher authorities. When the flood crisis struck, local and regional stakeholders retreated. Simultaneously, the crisis became an (inter)national media event, prompting the national government to step in and restore balance amidst the public outcry for more drainage. From the crisis, the observed one-sided approach to addressing European nature and water quality directives was a clear disaster. Alternative plans proposed by the Drainage Boards have the potential to deal with the double objectives better. While highly protected nature areas will remain protected and will continue to receive national funding, the danger is that intermediate areas that connect the designated nature areas will deteriorate and the habitat hotspots remain isolated.

In the eastern part of the Netherlands, all water authorities, provinces, municipalities, relevant NGOs, such as nature organizations and the farmers union, and drinking water companies agreed to develop and co-finance a working programme on fresh water supply in 2014. Projects like the *Vechtstromen* pilot fit perfectly in this programme. Another multilevel issue is the retreat of national government from nature and landscape policies. The Twente pilot area had been declared a "National Landscape" and an action plan was already developed when the national government stopped the policy programme leaving regional and local stakeholders to carry on with reduced support. In 2010, national government also cut nearly all support for the development of the National Ecological Network, a longstanding Dutch nature policy that aimed to halt this biodiversity loss by conserving nature, maintaining ecosystem function and service through connecting habitats. In the Twente pilot area, finding successful ways to upscale small-scale projects such that they cover larger areas, and also areas nearby, was difficult. This was not only true for farms, but also for nature conservation. Small fragments and discontinuities in nature areas are more vulnerable compared to areas that are linked and provide flora and fauna the ability to freely flow among areas. In this way, the ecological network approach was the ideal approach for the area. Now the Province of Overijssel had to step in to safeguard as much as possible, but could never do it to the same extent as would have been the case with more national support.

All in all, the *multilevel* interactions are quite varied between the regions studied. In most cases, water authorities have taken the lead in drought resilience management, though in the *Somerset* case, the primary actor is less apparent. In France, the national authorities have the lead in drought management, partly because of the centralized governance structure, and partly in response to the severe 2003 heat wave crisis. In the Somerset case the national authorities stepped in when the flood crisis became a big media event. In the two Dutch cases, national authorities withdrew from essential aspects of nature and landscape development. The European Union policies had strong indirect effects as a result of existing nature policies. Apart of the multilevel governance, the *multiscale* aspect of the measures and effects were also visible. In Eifel-Rur, the linked reservoirs spread out all over the area making measures at one spot relevant to other areas. In Vilaine, the reservoir level impacts areas far inland. In Twente, the challenge is to link the various scattered project areas where measures are taken into larger programmes that cover bigger areas.

12.5 Awareness on Nature Effects of Drought and the Public and Political Agenda

While nature protection not always directly affects human purposes, the awareness of threats to the integrity of the natural areas as ecosystems is not always self-evident. Sometimes the areas superficially look as nice as ever, even while droughts have undermined their viability.

In the higher sandy parts of *Flanders* that are dependent on groundwater and rain, drought-sensitive nature areas no doubt suffer from periods of desiccation, like adjacent areas in the Netherlands. However, it is remarkable how little attention this gets in Flanders. This phenomenon is highlighted in the Flemish Environment Report site (www.milieurapport.be—June 2015). The report does not mention nature effects of drought at all. While the Institute for Nature and Forest Research does give mention of its "nature indicators" with drought impacts as a keyword on its site (www.natuurindicatoren.be—June 2015), there is a lack of content on the specific impacts. In light of this, the fact that the Flemish Environment Agency, as part of DROP, is considering nature protection as one of the goals of drought adaptation is big advancement.

The Velpe and Dommel sub-pilots are within this drought-sensitive area. However, the activities there were more attuned to the impacts of agriculture (soil moisture) and calibration of the models. In the ongoing discussions on drought-related issues, environmental NGOs find it difficult to make their voices heard, due to the focus on agriculture and economic development.

While drought awareness is low in the *Eifel-Rur* area, drought can sometimes "piggy back" on more mainstream issues. For instance, protection of fish populations requires minimal water levels during periods of relative water scarcity which is also indirectly relevant to addressing drought. It is also a good objective for the national park authorities. Nature conservation groups, and to a lesser extent agriculture, are beginning to view the negative water balance during dry summers as a problem that needs to be addressed.

In the region of *Groot Salland*, most attention is focused on flood risk rather than drought. Still it would be unfair to state that drought risks are completely out of scope. In the 1980s, the Netherlands, established the term "desiccation of nature areas" as one of the "environmental themes". That said, despite this, very little was done to solve the problem in practice. Drought effects on nature remained a "soft" interest. When irrigation bans are issued in times of drought (often both for agriculture and garden watering) these measures have sufficient legitimacy. However, the relation with nature protection is not clear to the public in this case. Under the new irrigation policy, making sure that there is enough recharge of ground water levels quickly became a co-responsibility among farmers.

While drought effects on nature are not a big issue in *Vilaine*, nature protection as such is indeed an issue. A nature NGO named "Vivre les Marais" is promoting the responsible care for the Vilaine and Redon Natura 2000 wetlands. There is also a Natura 2000 plan for this in the works. The nature NGOs are involved in consultations, for example concerning the Water Plan. However, the NGOs involved have expressed feeling heard but not listened to. In the case of the wetlands, the main issue is not drought, but flooding. Here, the water level is controlled by the dam. In addition, drinking water and recreational (yachting) use provide incentives to raise reservoir water levels especially during normally dryer times, and often consequently flood inland marches at odd season. Beyond this scope, drought protection is not considered a serious issue for the region at the moment. Some awareness is evolving among the stakeholders, but its development is slow.

Even when in *Somerset* a number of subsequent droughts in 2010–2012 occurred, it was not always treated very seriously by the people in the region. Unlike nature organizations and some farmers who experience the variation of the water levels continuously, for most town dwellers droughts do not have immediately observed consequences as they do for nature. For town dwellers, drought is mostly correlated to fine weather. Nevertheless, all organized stakeholders were quite advanced to integrate drought aspects into their water-related climate adaptation plans. There was a proactive approach to drought management that was addressed across water supply, environment and even nature. After the big floods climate adaptation was reframed as recovery and mitigation from flooding, and reduced opportunities to include double-sided measures in the proposals to support the resilience of nature areas. Though the region is still recovering from the flood, there is still much reluctance to "piggy back" the flooding recovery with the issue of climate adaptation. This is likely remaining unchanged at least until the next big drought. Even then its consequences for nature should be cleverly communicated to a mostly urbanized society to increase the legitimacy of drought resilience measures.

While drought certainly is not a chief concern in the Twente region of *Vechtstromen*, awareness of drought risks for nature is widespread among institutional actors such as governments and NGOs. Water authorities and as well as province administration and the nature organizations also take the issue of drought seriously. This is less so among the broader public and parts of the agricultural community. The aim of pilot projects in the area is to raise awareness. Nature organizations in the region do regard desiccation as a serious problem, though the impact of nitrogen deposits is sometimes even more serious. Over time, the awareness is gradually growing, while the drought issue is repeatedly mentioned as one of the consequences of climate change. For the same reason occasional flooding is not interrupting this development: both sides of the coin are almost always mentioned together, and thus both floods and periods of drought contribute to the feeling that climate change is not a prediction anymore but an ongoing phenomenon.

The overview shows that the levels of *awareness* of climate change and drought impacts on nature are generally not very high and that they vary among different actors within each region. Among regions there is some variation in awareness, with the Twente region of Vechtstromen being most aware of drought and its impacts. The Somerset region is a close second in level of awareness. The floods events in 2013–2014 caused a great disruption, displacing drought awareness and placing concerns over drought in the rear. In spite of this, the *development over time* of drought awareness is showing a gradual increase. Consistent communication of the impacts of droughts and floods has helped to promote awareness and sustain dialogue and action surrounding both.

12.6   Conclusion: Highlighting the Main Issues
and Their Prospects

In an attempt to synthesize the broad variation observed across the drought and
nature policy context, the interplay of actors can be analyzed and summarized using
the three components of the contextual interaction theory, based on their motivations, cognitions and resources. Here, the motivation, cognitions and resource of
actors in the nature context are explored to analyze cross-cutting opportunities and
challenges for the drought policy context.

12.6.1   Motivation

Motivation refers to the goals and values, external pressure, and orientations that
drive the actor in a specific way. Among nature and conservation actors, the context
of water scarcity and drought produces highly varied motivations. In many cases,
the motivation of nature conservation stakeholders is triggered to a large extent by
their own goals and values. On the whole, both nature organizations and farmers are
aware of the implications of water scarcity and drought on nature areas, such as the
loss of flora and fauna in river systems, algae blooms and crop failures. Such
external pressures on a global environmental change scale serve as a primary
internal motivator among these actors.

   In addition, external governance pressures, such as the requirements set forth by
the Natura 2000 regulations provide a strong impetus to implement measures to
combat drought. Regulations, incentives and communications play a large role in
motivating action among stakeholders. The EU Directives in particular have created
pressures on relevant stakeholders (e.g. water boards) to devise collaborative
solutions to address water scarcity and drought. Despite such top-down pressures,
large regional disparities in resources and water use create cascading pressures at
the local level. Lack of resources tends to motivate local led initiatives to develop
their own innovative tools for drought-related problems.

   Despite strong economic interests, governance pressures can override them
depending on the strength of the regulation. For example, in the case of Somerset,
motivated actors from nature conservation organizations, including the Royal
Society for the Protection of Birds (RSPB) and the Somerset Wildlife Trust (SWT),
developed long-term visions for transforming the landscape from human-centred to
exclusively for nature by restoring open grassland landscapes. In the face of strong
economic interests, the benefits of a more resilient and robust landscape in the face
of extreme weather outweighed competing pressures. Such harmonization of initially competing motivations is at the heart of sustainable and integrative land and
water management as well as coping with extreme weather events, including
drought and flood (Robins 2014).

12.6.2 Cognitions

Cognitions refers to the observations of reality, the frames of reference, and the interpretations of the actors at hand. While motivations among actors tend to differ widely, all actors across case studies tend to share similar cognitions based on observed changes in the regional water balance. There are general observations among nature conservation actors that precipitation regimes are changing in unexpected ways and that these regime changes influence both flora and fauna habitats. In recent years, a large number of creeks have gone dry, leaving behind almost non-existent vegetation. Damages from droughts are increasingly affecting agricultural yields in rural areas, while cities and their urban infrastructure are also more and more at risk as a result of lower reservoirs.

The visibility of drought has increased in recent years, both in the farming and urban contexts as well, which in turn has contributed to widespread cognitive shifts among all nature conservation-related actors. More and more, stakeholders coherently and consistently agree that water scarcity and drought is, and will increasingly become, a problem. In response, strategies to combat drought impacts have already been developed into guidelines that keep water longer in the ground and in surface waters, employ water use efficiency schemes, and develop medium- and long-term possibilities for extra water transport. These strategies further contribute towards a more unified cognitions in nature arenas. As a result, it is expected that coherence among stakeholders will increase moving forward, with more experience with drought and water scarcity likely to develop into the future.

Nature-related regulations, including the WFD and the Habitats and Birds Directives, in addition to creating driving pressures, also contribute to shifting cognitions among relevant actors as drought protection measures become more ubiquitous.

More broadly, there is general cognition and awareness regarding the need for collaboration across a wide range of actors, including both public and private, at all levels. Because water scarcity and drought can impact both land, marine and freshwater ecosystems, connectivity is key to addressing the crises. Such a cognitive approach is particularly important in moving away from strictly legal incentives as the main motivators. In an effort to place less reliance on legal incentives, improving communication between actors and providing the tools and space for negotiating the relationship is the key. This approach allows for competing motivations to find more harmonious solutions before resorting to a priori legal recommendations.

12.6.3 Resources

Resources refer to the available capacity and power available internally and/or externally. On the whole, the resources dedicated to dealing with drought issues are limited. Amongst nature conservation actors, insufficient resources in terms of access to funds and general support are the most commonly cited barriers to engaging with water scarcity and drought issues. This is largely due to severe cuts in government funding, which have drastically reduced the amount of funding available for stakeholders. At present, in order to access funds, proposed measures in Germany require up to 80 % minimum co-financing by municipalities. When the remaining percentages are not achieved, the initiatives are either tabled or die. This not only leaves a large number of initiatives critical for addressing water scarcity and drought and other issues unfunded, but also it leaves the funds unused. This in turn has led to a large accumulation of financial resources that are available but not accessible.

In addition, actors within the nature conservation arena have increasingly resort to project acquisition (such as LIFE or INTERREG) as well as available funding at the state level for specific projects, such Natura 2000 monitoring. An alternative approach, particularly applied in governing nature reserves, that has emerged are public–private partnerships (PPPs). PPPs are increasingly the trend to maintain conservation areas and the ecosystem services which they provide. While the trend is promising, it also underscores the decreasing flexibility of governance systems to address new policy issues such as drought and water scarcity.

Simultaneously, there is a lack of resources in terms of the competencies among nature conservation actors that has limited them to tackling issues of drought. Due to a lack of power, and also agency, they often lack the flexibility that other actors, such as the water boards, possess.

Lastly, there exist limited resources in the legal sense as well. Competing interests between agriculture and nature areas have been at the heart of legal tensions as impacts of drought are felt on both sides. Though measures to encourage water efficiency schemes have helped to divert tensions, there is a critical need for more integrated land and water management perspectives in order to avoid resorting to legal tools.

The interplay of motivations, cognitions and resources in the case study processes shows that keeping the nature perspective influential is sometimes a difficult task. Even so, in many areas there are actors that see it as a vulnerable aspect of drought resilience and are prepared to protect it.

Reference

Robins M (2014) Drowning out nature on the levels? ECOS 35(1):27–30

Chapter 13
Towards a Drought Policy in North-West European Regions?

Corinne Larrue, Nanny Bressers and Hans Bressers

13.1 Introduction

As presented in the previous chapters, to enhance the preparedness of NW European regions for periods of drought and water scarcity, the governance team used a governance assessment tool (GAT) to reveal the 'essence' of drought adaptation and governance in the six NW European regions investigated (see Chap. 3). We should remember that this governance assessment has been developed by social scientists with the help of practice partners (project partners from the region, such as water authorities and county councils) and other governmental and non-governmental stakeholders. This inclusion of practice partners has allowed a continuous iteration between science and practice, as well as access to regional stakeholders for interviews; in addition, it ensured an even representation of the relevant stakeholders. The contacts and networks of the practice partners facilitated the exchange with these regional stakeholders.

This 'Governance Assessment Tool' is composed of a 'matrix' style model that consists of five elements (levels and scales, actors and networks, perceptions of the problem and goal ambitions, strategies and instruments, and responsibilities and resources for implementation) and four criteria (extent, coherence, flexibility and intensity), producing a matrix of 20 cells. This model was used to diagnose the regional setting and to formulate regional roadmaps to optimize regional settings. As presented in the conclusion of the previous chapters, a qualitative evaluation has been performed for each region. For each case, the evaluation of the drought

C. Larrue (✉)
Paris School of Planning, University of Paris EST, Champs-sur-Marne, France
e-mail: corinne.larrue@u-pec.fr

H. Bressers
CSTM, University of Twente, Enschede, The Netherlands

N. Bressers
Water Authority Vechtstromen, Almelo, The Netherlands

© The Author(s) 2016
H. Bressers et al. (eds.), *Governance for Drought Resilience*,
DOI 10.1007/978-3-319-29671-5_13

governance context has been summarized by assessing each of the 20 cells through a graphical visualization that shows the matrix with colours ('score cards'); these indicate the value of each cell (restrictive, neutral or supportive context).

Based on the conclusions of the cross-cutting perspective chapters and case study chapters, it is useful to propose a comparative approach of the drought governance context in the six regions studied. This comparison allows us to outline general trends that emerged from each of the cases and to show possible specificities of the regions studied. This comparison also allows us to analyse the specificity of each of the cross-cutting issues (i.e. nature, fresh water and agriculture) to note the contexts that facilitate or prevent a better drought governance context.

However, transferring the richness of the data gathered by numerous documents and interviews into more condensed layers of summary and ultimately into an overview has both positive and negative aspects. On the one hand, this transfer is necessary to enable a comparative analysis between several cases; however, on the other hand, the summary should not hide essential observations that provide evidence for the scores. Thus, one should always remember that such a summary of summaries is a derivative of a much richer set of observations and their interpretation. In addition, the matrices have been implemented independently by different leading authors. In order to overcome the fact that certain authors differ slightly in their 'judgments', several meetings with all the analysts have been held in order to reach a common agreement upon these assessments. Hence, for the comparison of the assessment of the different case studies, we used a greater amount of the written text for the assessments and did not only use the comparison of the matrices. However, to illustrate our comments in this text we based our comparative statements on the coloured matrix stemming from the regional case studies.

This chapter is devoted to concluding remarks. It will first present certain overarching observations related to case study results (Sect. 13.2) and to the three cross-cutting perspectives presented in the previous chapters (Sect. 13.3). In conclusion, we will then outline a few recommendations (Sect. 13.4).

13.2   How Governance Can Be Characterized in Each Region?

The outcomes of the analysis of the drought-related water governance issues in the NW European regions involved in the DROP project can be summarized by the three following main points:

- A low level of awareness exists, as regards the drought issue, creating a poor context for responsibilities and resources, and leading to a very low level of intensity of drought-related actions
- However, an effective water governance, particularly for networks of actors, and their involvement at different levels and scales exists in all regions
- Although variable according to the region, there is a low level of flexibility in the governance context

13.2.1 A Low Level of Awareness as Regards the Drought Issue

The problem with perceptions and goal ambitions is that this is the dimension in which the governance context does not favour drought policy. The four criteria of this dimension are either neutral (yellow) or restrictive (red) in most of the studied regions, particularly where the intensity is concerned. This is shown in Table 13.1.

In fact, in most of the studied regions and primarily because of the traditionally wet situation of the NW regions, many actors involved in water governance are not aware of the potential drought situation or do not see it as a priority. These actors are much more preoccupied with floods. Additionally, in the Somerset case, in which the awareness of drought impacts was high at the beginning of the project, the flood event that occurred during the course of the DROP project changed the minds of those involved and allowed them to forget the drought issue for some period.

In nearly all cases, the intensity of problem perceptions as well as of goal ambitions is the worst dimension for the drought governance context. In several cases drought issues were introduced during the interviews with the governance team.

This low level of drought awareness results in a low-intensity assessment of all of the dimensions of the governance context (Table 13.2). If we consider the intensity criteria for all the dimensions in each region, we can assess that the intensity quality is either restrictive (red) or neutral (yellow) in all of the regions and

Table 13.1 Assessment of problem perceptions and goal ambitions criteria in the six regions

	Extent	Coherence	Flexibility	Intensity
SALLAND				
FLANDERS				
TWENTE				
SOMERSET				
VILAINE				
EIFEL RUR				

Table 13.2 Intensity assessment for all of the dimensions in all of the regions

Intensity	Salland	Flanders	Twente	Somerset	Vilaine	Eifel Rur
Levels and scales						
Actors and networks						
Problem perspectives and goal ambitions						
Strategies and instruments						
Responsibilities and resources						

almost all of the dimensions. Taking into account the outcome of the regional case study analysis we can relate low level of drought awareness and low level of intensity of the drought governance context in NW regions.

Next to the dimension of problem perception and goal ambitions, also the dimension of responsibilities and resources for implementation is problematic. It generally scored low on coherence, flexibility and intensity, though not on extent.

In most of the regions studied, the drought issue is not completely out of sight. The actors interviewed are aware of the potential occurrence of such a drought situation. Therefore, in most cases, the governance team members positively assessed the extent of the responsibilities and resources or the strategies and instrument dimensions. However, this does not always imply a true involvement of the stakeholders, preventing them from developing a coherent policy in this area. The anticipation capacity throughout all of the regions is limited to a few measures, the relevance of which is easily challenged if other more urgent problems arise, as in the Somerset case.

13.2.2 Effective Water Governance as Regards Actors and Their Networks in All of the Regions

In contrast to the situation of awareness and intensity, we observed a much better drought governance context as regards the actors and networks dimension as well as the levels and scales dimension (see Table 13.3 for the actors and networks dimension). In all of the regions studied, the actors involved at different decision levels are mobilized, which constitutes a context that is particularly favourable to the establishment of a drought policy that integrates these different levels. Most of the qualities in the actor and network dimension have been assessed by the governance team members as supportive (green) or neutral (yellow), though even here intensity remains the weakest part.

These conclusions can be related to the existing water governance systems in most regions. In all countries, water governance is relatively effective because a water policy has already been implemented since the 60s. This aspect of being a

Table 13.3 Actors and networks criteria in all of the regions

	Extent	Coherence	Flexibility	Intensity
SALLAND				
FLANDERS				
TWENTE				
SOMERSET				
VILAINE				
EIFEL RUR				

Table 13.4 Coherence assessment for all of the dimensions in all of the regions

Coherence	Salland	Flanders	Twente	Somerset	Vilaine	Eifel Rur
Levels and scales						
Actors and networks						
Problem perspectives and						
Strategies and						
Responsibilities and resources						

relatively established sector has been recently reinforced by the implementation of the Water Framework Directive (WFD), which imposes a multilevel water policy through the formulation of a district management plan.

More precisely, if we consider the coherence criteria for all the dimensions in each region, we can assess a supportive governance context as regards this quality. Most of the cells have been assessed supportive (green) or neutral (yellow). However, this is essentially true for both of the dimensions, 'levels and scales' and 'actors and networks'. When addressing issues that are more closely related to drought as regards responsibilities or strategies, the coherence appears to be much less evident in each of the regions studied (Table 13.4). This evaluation reflects the fact that the drought issue is not truly at stake for several water actors.

Moreover, it stems from the more detailed regional case study reports that where it exists, the governance consistency is mainly due to strong interrelationships between actors based on mutual trust.

More generally, water governance implemented within each of the regions produced interactive knowledge between actors: most of them met often and know well each other's perspectives and have developed a common knowledge around water issues. Even if their position can be conflictual, they share a mutual interest in water management, which can help the future formulation and implementation of drought policy.

13.2.3 Although Variable According to the Region, There Is a Low Level of Flexibility of the Governance Context

Considering the flexibility criteria for all the dimensions in each region, the analysis highlights a more neutral context (Table 13.5). Although different actors are mobilized at different decision-making levels, the system of interactions between the players apparently does not allow sufficient flexibility to enable the decision-making system to easily incorporate new issues, such as drought or water scarcity. The governance context is not truly prepared to address the water scarcity and drought issue and to integrate it as an important issue.

Table 13.5 The flexibility assessment for all the dimension in each of the regions

Flexibility	Salland	Flanders	Twente	Somerset	Vilaine	Eifel Rur
Levels and scales					███	
Actors and networks						
Problem perspectives and goal ambitions				███		
Strategies and instruments						
Responsibilities and resources						

However, the flexibility of governance may be more constrained by the institutional context of each country or region than by the level of consciousness. We can then observe that the Flanders case as well as the Eifel-Rur case appear to have the most neutral context as regards flexibility (all of the dimensions are assessed as neutral as pointed out in Table 13.5). This assessment can be related to the lack of flexibility in general which has been pointed out in the case study chapters about these two regions.

In sum, the implementation of the GAT leads to the conclusion that the governance context for drought resilience policies and measures in most of the regions studied can be regarded to currently be 'intermediate'. This tool does not conclude to a clear positive or negative picture of the drought and water scarcity governance context in those NW regions. The governance circumstances appear to be half capable of providing a favourable context in terms of the actors and decision levels involved in all of the regions, but do not provide a really favourable context to develop and implement a coherent drought policy.

13.3   Outcome of the Analysis: A Cross-cutting Perspective

Drought or water scarcity situations either during a short period or as a more structural pattern, leads to readdress the issue of allocation of water uses and the related water user rights. In NW European regions this water management issue is changing: from how to better allocate water between sectors towards how to minimize impacts from one user or stakeholder to another, trying to better combine uses. However, for the six regions studied this question is only partially at stake until now, due to the low level of drought awareness as pointed above.

Indeed while we can witness a beginning trend to view the negative water balance during dry summers as a problem that needs to be addressed, very little was done up to now to solve the problem in practice.

Stemming from the cross-cutting analysis presented above, three mains issues can be pointed out in order to characterize the way the sectors' needs are taken into account.

- Water governance that in general gives more weight to representatives of economic interests than to environmental ones
- A hierarchy as regards water uses in case of water scarcity that favours drinking and service water supplies
- Contrasting initiatives which try to better take into account drought in all sectors

13.3.1   Water Governance Gives More Weight to Representatives of Economic Interests Than to Environmental Ones

Together with traditional water users (water supply, industries, etc.) the agricultural sector relies upon water rights that it holds from the past. It is thus hard to take these rights and to redistribute them among new sectors as nature-related ones.

More precisely, due to the relative importance of the agricultural sector in the regional economy of the studied regions, the agriculture production influences water governance. The governance of drought and water scarcity reserves an important listening ear towards representatives of agricultural sectors in all the regions. Moreover, this economic sector proved to be well organized and to operate with a high level of interactive capacities in all the regions.

One can notice that multilevel interactions are quite varied between the regions studied. In most cases, water authorities have taken the lead in drought resilience management, but the involvement of other stakeholders is not balanced: environmental NGOs find it difficult to make their voices heard, due to the focus on agriculture and economic development (tourism, urbanization, etc.).

However in some regions like the ones studied in the Netherlands, it is worth noting that agriculture is not always ranked at the highest priority level. In that country, part of surface water needs to be flushed to prevent salt intrusion in the low-lying western parts of the country, which implies a lower availability for agricultural use in the east part, and within the displacement chain, which outlines the priorities of competing water uses during dry periods, agricultural production has a low priority. This hierarchy between sectors must thus be analysed through the lenses of geographical and sociopolitical context.

It has been pointed out in the nature cross-cutting chapter that in each region, the water system has been modified in the past, either for purposes of flood protection or freshwater delivery or agricultural efficiency or all at the same time. All of these interventions had side effects on other services of the water system that were often not recognized when the interventions took place. These modifications question the level of 'naturality' to be taken into account which does not help to reallocate water rights to 'nature-related stakes'.

13.3.2   A Hierarchy as Regards Water Uses in Case of Water Scarcity that Favours Water Supplies

The three cross-cutting chapters clearly show that the impact of drought on water supplies is generally taken into account than the two other issues: agriculture and nature. In all of the cases studied, the priority is given to human water uses when the resource becomes scarce, even if we witness a growing sensibility, which tends to question this hierarchy especially in the Netherlands. In all the regions, freshwater availability for drinking water provision is usually ranked as a priority stake.

Moreover, the regional analysis pointed out that disruption in freshwater delivery is considered as a management mistake. That explains the difficulty to question the hierarchy already settled in favour of freshwater supply.

Last, it is worth noting that the impact of drought is not only a quantitative issue. For freshwater supply it is mostly a qualitative one. As stated above, during drought episodes, water quality in lakes and reservoirs generally shows deterioration due to less dilution, particularly for nutrients and salinity which oblige to increase water treatment for drinking water production. This is why in some regions the main focus is given to be better prepared for crisis management.

13.3.3   Contrasting Initiatives Which Try to Better Take into Account Drought in All Sectors

While our first two observations are somewhat pessimistic that are also numerous initiatives that have been pointed out in the six regional case study chapters which cannot all be reported here. We can stress here the main trends along four main lines:

- **Awareness**: even if awareness proved to be low in each region, gradually the agricultural sector becomes more sensitive to drought risk as a result of past events. Moreover, it has been stressed in some cases that the fundamental nature of drinking water provision and freshwater availability may be the leading entryway to the development of drought risk awareness and drought adaptation measures.
- **Knowledge**: a better knowledge about agricultural and individual water uses is recognized in all the regions as a necessity, and new tools are implemented in order to strengthen this knowledge capacity. To that respect, the developed web platform (www.water.be) with the modelling results in Flanders can increase the awareness for the problem, if it is used as an information channel, e.g. by farmers. Increased communication of risk to actors which have underdeveloped risk perception is then recommended. For these activities the developed scientific model can be used, but the more technical approach should be combined

with a more interactive approach, e.g. with showcases on local level together with farmers.

- **Engineering**: in most regions there is a need for additional buffering capacity by enlarging the infrastructure interconnectivity among catchments. Connectivity appears to be the key to addressing the crises. In some regions, water saving technics in agriculture and other sectors begin to be developed.
- **Planning**: there is a critical need for more integrated land and water management perspectives. The mobilization of new water resources, through the building of water retention basins, for instance must not be considered as a paramount unique solution and must be integrated within a comprehensive drought plan.

13.4 Conclusions and Recommendations Stemming from the Implementation of the GAT

In conclusion of this presentation of the governance assessment's main outcomes, we can generally state that the context of governance could be greatly improved by an awareness of the importance of drought conditions and a greater focus on its prevention.

Forty recommendations have been made to the regional practice partners by the governance team members. We will present some of our more general observations and recommendations, which partly stem from regional observations, but which have relevance and transferability to other regions in North-west Europe.

13.4.1 Continuous Focus on Realizing Awareness Is Needed

The main and first recommendation to be stressed is a focus on raising the awareness of drought and water scarcity in all of the NW regions.

Across the areas studied, we found that the problem was that the awareness among land owners and the general public, and thus many politicians, remains low. This lack of awareness restricts the selection of forceful interventions to increase drought resilience and occasionally makes it more difficult to practically realize the measures chosen.

Based on the governance team visits, the discussions with the water authorities and with many other stakeholders, and the results of the governance assessment, it was possible to achieve certain major recommendations regarding this central issue, which is the awareness and strengthening of drought and water scarcity issues' position on public and political agendas in the various countries.

We can distinguish three major strategies for pushing the position of the drought issue that is still experienced by many as a second-order issue.

(1) Aiming to place drought and water scarcity on the public and political agendas on their own, as independent problems; for instance, by providing continuous information to the public, such as in Flanders, on the agency's website or by directly addressing national water planners with a broad coalition of stakeholders, such as in the Netherlands' Delta programme process.

(2) Addressing drought by 'piggybacking' other issues, i.e. including drought-relevant measures in different planning initiatives and ensuring the coherence of plans with drought objectives.

(3) Preparing a ready-to-implement strategy for when a drought event makes the topic climb the agenda and receive political attention, resulting in a call for action.

The careful application of a combination of these strategies leads to the best positioning of drought issues and aligns them more closely with the already recognized importance of flood risks.

More generally, the issue of drought can be related to the issue of climate change: better knowledge of the regional effects of climate change on water availability can be a first step to improving visibility and understanding of the problem, even if this might be obviously not true everywhere, that it will not be sufficient as such.

13.4.2 Preparation and Implementation of Water Demand Management

Most measures involve distributing the available water and decreasing water scarcity during dry periods to make the areas more resilient by improving their water buffering capacity. Until now, measures oriented towards water demand have been less common. However, in the future, they might need to become a more common part of the drought resilience strategy, even in several areas in water-rich North-west Europe. This implies a current need for the collection of data on water rights, the following of water uses, and a review of water prices. Policy measures and instruments should generate incentives for use reduction that are now often absent, as water is still regarded as a free commodity rather than as a scarce resource. Thus, fostering the mainstreaming of drought risk and drought preparedness into private actors' activities is important.

Collaboration with farmers proved to be very important in most cases in the DROP project, not only in the two cases in which the pilot project was explicitly addressing agriculture. Increasing synergies with agriculture, e.g. through farmer advisory services or the inclusion of farmers unions in the design and implementation of measures, seems to be a prerequisite for successful demand management.

The management of expectations might be equally important. As long as the abundance of water in North-west Europe is taken for granted by water users, the implicit responsibility to protect the water supply is placed by the users on the water authorities. Although water supply should remain a public task, it does not follow that the water authorities and taxpayers' money should accommodate increased vulnerability for shortages or new economic activity requiring extra fresh water or water of a specific quality. Some investments might not be wise in areas prone to drought. Openly discussing the limits of public responsibility might increase the awareness and ownership of preparatory measures of such water users.

13.4.3 The Need for an Increased Integration of Flood and Drought Management

In our project, the UK Somerset case is a clear example of a situation in which, after several years of droughts, the large 2013–2014 flood disturbed the balance between drought and flood measures and proved that both are sides of the same climate change adaptation coin. It is essential to consider surplus water events when taking drought resilience measures, and vice versa. Recognizing the need to address the impact of floods, while acknowledging that there is also a very real threat of water scarcity in North-west Europe, changes the range of strategies and instruments that could be used to effectively mitigate variability and extremes. This more closely aligned approach of different forms of water management draws together a range of lessons for more the effective governance of climate change adaptation across the whole of North-west Europe. We need strategic governance approaches that are focused on adaptation and resilience of the entire water system rather than on crisis management of extreme events.

13.4.4 Variety Requires Tailored Action

Each of our six regional reports contains specific background information, analyses of governance conditions, and some recommendations on how to deal with the regional water governance situation from a drought perspective. The recommendations are partially based on a comparison of the specific region's context with the Governance Assessment Team members' knowledge of other water management systems, including a comparative analysis of the other five regions studied in DROP.

In the six regions studied, there are wide varieties of drought measures implemented—involving inter alia and drought prediction models as well as building infrastructure for improved water level management, natural water retention measures, and farmer-targeted assistance to improve irrigation practices. This variety reflects the need for tailored action due to the variety of natural situations in the different regions of North-west Europe. While increased insight and data processing are needed to better understand the dynamics of the water system in regard to drought issues, the best measures are highly dependent on the geohydrological situation and structure of water demand. However, the governance context also has a clear influence on the development of habitual approaches in policy-making and implementation. Some of the variety is not so much the result of physical conditions, but more so of governance settings.

These four main recommendations, drawn from the case studies and cross-cutting studies, may help regional actors to formulate and implement means able to address regional drought adaptation.